Praise for *Football Analytics with Python & R*

Grounded in clarity and real-world mentorship, Eric and Richard demystify football analytics for the masses, both novices and practitioners alike. This foundational guide transforms complex data into accessible wisdom, making it a must-read for anyone eager to better understand the game.

—*John Park, Director of Strategic Football Operations,*
Dallas Cowboys

This book is the perfect mix of applicable storytelling and use cases within football analytics. It's an ideal resource for setting the foundation of a career in football analytics, with specific coding examples and broad coding instruction. Anyone can pick this up and learn to code with football data.

—*John Taormina, Director of Football Strategy, TruMedia,*
former Atlanta Falcons Director of Football Data & Analytics

The case study-driven approach makes it easy to understand how to approach the most common tasks in football analytics. I love how the book is filled with data science and football lore. It's a must-read for anyone interested in exploring football data.

—*Richie Cotton, Data Evangelist at DataCamp*

A practical guide for learning, implementing, and generating insight from meaningful analytics within the world of football. Fantastic read for sports enthusiasts and data-driven professionals alike.

—*John Oliver, data scientist*

This is a great reference to learn how data science is applied to football analytics. The examples teach a wide range of visualization, data wrangling, and modeling techniques using real-world football data.

—*Ryan Day, advanced data scientist*

One of the rare books out there written by data science educators that also work in the industry. Football analytics in both Python and R throughout, too! Excellent, thoughtful examples as well.

—*Dr. Chester Ismay, educator, data scientist, and consultant*

Football Analytics with Python & R
Learning Data Science Through the Lens of Sports

Eric A. Eager and Richard A. Erickson

Beijing · Boston · Farnham · Sebastopol · Tokyo

Football Analytics with Python & R

by Eric A. Eager and Richard A. Erickson

Published by O'Reilly Media, Inc., 1005 Gravenstein Highway North, Sebastopol, CA 95472.

O'Reilly books may be purchased for educational, business, or sales promotional use. Online editions are also available for most titles (*https://oreilly.com*). For more information, contact our corporate/institutional sales department: 800-998-9938 or *corporate@oreilly.com*.

Acquisitions Editor: Michelle Smith	**Indexer:** nSight, Inc.
Development Editor: Corbin Collins	**Interior Designer:** David Futato
Production Editor: Aleeya Rahman	**Cover Designer:** Karen Montgomery
Copyeditor: Sharon Wilkey	**Illustrator:** Kate Dullea
Proofreader: Amnet Systems LLC	

August 2023:   First Edition

Revision History for the First Edition

2023-08-16:   First Release

See *https://oreilly.com/catalog/errata.csp?isbn=9781492099628* for release details.

978-1-492-09962-8

[LSI]

Table of Contents

Preface

Life is about stories. For many, these stories revolve around sports, and football specifically. A lot of us remember playing games in the backyard with our family and friends, while some were lucky enough to play those games under the lights and in front of fans. We love our heroes and loathe our rivals. We tell these stories to help us understand the past and try to predict the future. Football analytics, at its core, allows us to use information to tell more accurate stories about the game we love. They give us ammunition for some of the game's most vexing questions.

For example, do the most championships make the best quarterback? Or the best passing statistics? What do the "best passing statistics" even mean? Are there things that players do that transcend the math? If so, are they simply transcending our current understanding of the game mathematically, only to be shown as brilliant once we have new and better information? Who should your favorite team draft if it is trying to win this year? What about in three years from now? Who should your grandmother draft in fantasy football? Which team should you bet on during your layover in Las Vegas during a football Sunday?

Following the success of the *Moneyball* approach in baseball (captured by Michael Lewis in the 2003 book published by W. W. Norton & Company), people now increasingly use math and statistics to help them properly formulate and answer questions like these. While football, an interconnected game of 22 players on a 100-yard field, might seem less amenable to statistical analysis than baseball, a game with many independent battles between pitcher and batter, folks have made substantial progress toward understanding and changing the game for the better.

Our aim is to get you started toward understanding how statistical analysis may also be applied to football. Hopefully, some of you will take what you learn from this book and contribute to the great data revolution of the game we all love. Each chapter of this book focuses on a problem in American football, while covering skills to answer the problem.

- Chapter 1, "Football Analytics", provides an overview of football analytics to this date, including problems that have been solved within the last few years. It then explores publicly available, play-by-play data to measure aggressiveness by National Football League (NFL) quarterbacks by looking at their average depth of target (aDOT).

- Chapter 2, "Exploratory Data Analysis: Stable Versus Unstable Quarterback Statistics", introduces exploratory data analysis (EDA) to examine which subset of quarterback passing data—short passing or long passing—is more stable year to year and how to use such analyses to look at regression candidates year to year or even week to week.

- Chapter 3, "Simple Linear Regression: Rushing Yards Over Expected", uses linear regression to normalize rushing data for NFL ball carriers. Normalizing data helps us adjust for context, such as the number of yards a team needs for a first down, which can affect the raw data outputs that a player produces.

- Chapter 4, "Multiple Regression: Rushing Yards Over Expected", expands upon the work in Chapter 3 to include more variables to normalize. For example, down and distance both affect expectation for a ball carrier, and hence both should be included in a model of rushing yards. You'll then determine whether such a normalization adds to stability in rushing data. This chapter also explores the question "Do running backs matter?"

- Chapter 5, "Generalized Linear Models: Completion Percentage over Expected", illustrates how to use logistic regression to model quarterback completion percentage.

- Chapter 6, "Using Data Science for Sports Betting: Poisson Regression and Passing Touchdowns", shows how using Poisson regression can help us model game outcomes, and how those models apply to the betting markets.

- Chapter 7, "Web Scraping: Obtaining and Analyzing Draft Picks", uses web-scraping techniques to obtain NFL Draft data since the start of the millennium. You'll then analyze whether any teams are better or worse than expected when it comes to picking players, after adjusting for the expectation of the pick.

- Chapter 8, "Principal Component Analysis and Clustering: Player Attributes", uses principal component analysis (PCA) and clustering to analyze NFL Scouting Combine data to determine player types via unsupervised learning.

- Chapter 9, "Advanced Tools and Next Steps", describes advanced tools for those of you wanting to take your analytics game to the next level.

In general, these chapters build upon one another because tools shown in one chapter may be used later. The book also includes three appendixes:

- Appendix A, "Python and R Basics", introduces Python and R to those of you who are new to the programs and need guidance obtaining them.
- Appendix B, "Summary Statistics and Data Wrangling: Passing the Ball", introduces summary statistics and data wrangling by using an example to demonstrate passing yards.
- Appendix C, "Data-Wrangling Fundamentals", provides an overview of more data-wrangling skills.

We have also included a Glossary at the end of the book to define terms.

We use case studies as the focus of this book. Table P-1 lists the case studies by chapter and the skills each case study covers.

Table P-1. Case studies in this book

Case study	Technique	Location
Motivating example: Home field advantage	Example framing a problem	"A Football Example" on page xiv
Pass depth for quarterbacks	Obtaining NFL data in R and Python	"Example Data: Who Throws Deep?" on page 10
Passing yards across seasons	Introduction to stability analysis with EDA	"Player-Level Stability of Passing Yards per Attempt" on page 37
Predictor of rushing yards	Building a simple model to estimate rushing yards over expected (RYOE)	"Who Was the Best in RYOE?" on page 69
Multiple predictors of rushing yards	Building a multiple regression to estimate RYOE	"Applying Multiple Linear Regression" on page 94
Pass completion percentage	Using a logistic regression to estimate pass completion percentage	"GLM Application to Completion Percentage" on page 122
Betting on propositions (or props)	Using a Poisson regression to understand betting	"Application of Poisson Regression: Prop Markets" on page 140
Quantifying the Jets/Colts 2018 trade	Evaluating a draft trade	"The Jets/Colts 2018 Trade Evaluated" on page 192
Evaluating all teams drafting	Comparing drafting outcomes across all NFL teams	"Are Some Teams Better at Drafting Players Than Others?" on page 194
Player NFL Scouting Combine attributes	Using multivariate methods to classify player attributes	"Clustering Combine Data" on page 230

Who This Book Is For

Our book has two target audiences. First, we wrote the book for people who want to learn about football analytics by *doing* football analytics. We share examples and exercises that help you work through the problems you'd face. Throughout these examples and exercises, we show you how we think about football data and then how to analyze the data. You might be a fan who wants to know more about your team, a fantasy football player, somebody who cares about which teams win each week, or an aspiring football data analyst. Second, we wrote this book for people who want an introduction to data science but do not want to learn from classic datasets such as flower measurements from the 1930s or *Titanic* survivorship tables from 1912. Even if you will be applying data science to widgets at work, at least you can learn using an enjoyable topic like American football.

We assume you have a high school background in math but are maybe a bit rusty (that is to say, you've completed a precalculus course). You might be a high school student or somebody who has not had a math course in 30 years. We'll explain concepts as we go. We also focus on helping you see how football can supply fun math story problems. Our book will help you understand some of the basic skills used daily by football analysts. For fans, this will likely be enough data science skills. For the aspiring football analyst, we hope that our book serves as a springboard for your dreams and lifelong learning.

To help you learn, this book uses public data. This allows you to re-create all our analyses as well as update the datasets for future seasons. For example, we use only data through the 2022 season because this was the last completed season before we finished writing the book. However, the tools we teach you will let you update our examples to include future years. We also show all the data-wrangling methods so that you can see how we format data. Although somewhat tedious at times, learning how to work with data will ultimately give you more freedom: you will not be dependent on others for clean data.

Who This Book Is Not For

We wrote this book for beginners and have included appendixes for people with minimal-to-no prior programming experience. People who have extensive experience with statistics and programming in R or Python would likely not benefit from this book (other than by seeing the kind of introductory problems that exist in football analytics). Instead, they should move on to more advanced books, such as *R for Data Science*, 2nd edition by Hadley Wickham et al. (O'Reilly, 2023) to learn more about R, or *Python for Data Analysis*, 3rd edition by Wes McKinney (O'Reilly, 2022) to learn more about Python. Or maybe you want to move into more advanced books on topics we touch upon in this book, such as multivariate statistics, regression analysis, or the Posit Shiny application.

We focus on simple examples rather than complex analysis. Likewise, we focus on simpler, easier-to-understand code rather than the most computationally efficient code. We seek to help you get started quickly and connect with real-world data. To use a quote often attributed to Antoine de Saint-Exupéry:

> If you wish to build a ship, do not divide the men into teams and send them to the forest to cut wood. Instead, teach them to long for the vast and endless sea.

Thus, we seek to quickly connect you to football data, hoping this connection will inspire and encourage you to continue learning tools in greater depth.

How We Think About Data and How to Use This Book

We encourage you to work through this book by not only reading the code but also running the code, completing the exercises, and then asking your own football questions with data. Besides working through our examples, feel free to add in your own questions and create your own ideas. Consider creating a blog or GitHub page to showcase your new skills or share what you learn. Work through the book with a friend or two. Help each other when you get stuck and talk about what you find with data. The last step is especially important. We regularly think about and fine-tune how we share our datasets as we work as professional data scientists.

In our day jobs, we help people make decisions using data. In this book, we seek to share our tools and thought processes with you. Our formal academic training covered mathematics and statistics, but we did not truly develop our data science skills until we were required to analyze messy, ecological, and environmental data from the natural world. We had to clean, merge, and wrangle untidy data, all while figuring out what to do with gaps in the information available to us. And then we had to try to explain the meaning hidden within those messy datasets.

During the middle of the last decade, Eric starting applying his skills to football, first as a writer and then as an analyst, for a company called Pro Football Focus (PFF). Eventually, he left academia to join PFF full time, helping run its first data science group. During his time with PFF, he worked with all 32 NFL teams and over 130 college teams, before moving to his new role at SumerSports. Meanwhile, Richard continues to work with ecological data and helps people make decisions with this data—for example, how many fish need to be harvested from where in order to control an undesired species?

Although we both have advanced degrees, the ability to think clearly and explore data is more important than formal education. According to a quote attributed to Albert Einstein, "Imagination is more important than knowledge." We think this holds true for football analytics as well. Asking the right question and finding a good enough answer is more important than what you currently know. Daily, we see how quantitative tools help us to expand our imagination by increasing our ability to look at and think about different questions and datasets. Thus, we are required to imagine important questions to guide our use of analytics.

A Football Example

Let's say we want to know if the Green Bay Packers have a home-field advantage. Perhaps we disagree with a friend over whether the Frozen Tundra truly is the advantage that everyone says it is, or if fans are wasting their hard-earned money on an outsized chance of getting frostbite. Conceptually, we take the following steps:

1. We find data to help us answer our question.
2. We wrangle the data into a format to help us answer our question.
3. We explore the data by plotting and calculating summary statistics.
4. We fit a model to help us quantify and confirm our observations.
5. Lastly, and most importantly, we share our results (optionally, but possibly most importantly, we settle the wagers to the "questions" we answered with data).

For the Packers home-field advantage example, here are our concrete steps:

1. We use the `nflreadr` package (*https://nflreadr.nflverse.com*) to obtain data, which is freely available to use.
2. We wrangle the data to give us a score differential for each game.
3. We create a plot such as that in Figure P-1 to help visualize the data.
4. We use a model to estimate the average difference point differential for home and away games.
5. We share our results with our friends who were debating this topic with us.

Given the data, the Packers typically have a point differential of two points higher at home compared to being on the road. This is in line with where home-field advantage is assumed to be across the league, although this number has evolved substantially over time.

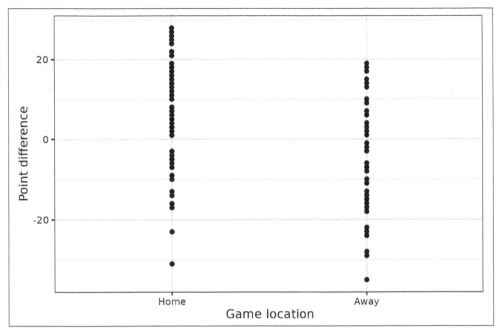

Figure P-1. Green Bay score differential in home and away games from 2016 to 2022

Hopefully, this observation raises more questions for you. For example, how much variability exists around this estimate? What happens if you throw out 2020, when fans were not allowed to attend games at the Packers home stadium? Does home-field advantage affect the first half or second half of games more? How do you adjust for quality of play and schedule differential? How does travel distance affect home-field advantage? Familiarity? Both (since familiarity and proximity are related)? How does weather differential affect home-field advantage? How does home-field advantage affect winning games compared to losing games?

Let's quickly look at homefield advantage during games won by the Packers compared to games lost. Look at Figure P-2. Summaries for the data in this figure show that the Packers lose by one point more on the road compared to home games, but they win at home by almost six points more than they do on the road.

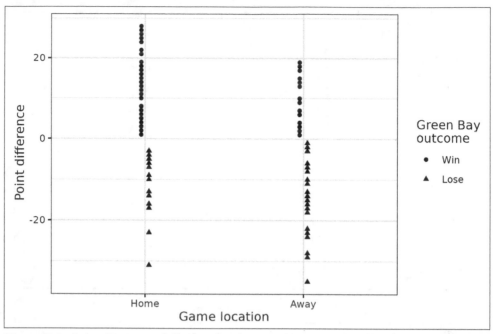

Figure P-2. Green Bay score differential in home and away games from 2016 to 2022 for winning and losing games

That being said, the Green Bay Packers are a good team and will likely have a positive point differential no matter what. The question of home-field advantage is not a trivial one and has perplexed analysts and bettors alike for decades. Hopefully, this example spurs more questions for you. If so, you're reading the right book.

We cover step 1, obtaining football data, in Chapter 1. We cover step 2, exploring data, in Chapter 2. We cover step 3, data wrangling, in case studies throughout the book as well as in Appendixes B and C. We cover step 4 with basic statistics in Chapter 2 and Appendix B and then cover models in Chapters 3 through 8. We cover step 5 throughout various chapters as we describe what we have found. Lastly, we round out the book with Chapter 9, which describes some of the advanced tools we use daily.

What You Will Learn from Our Book

We have included materials in this book to help you start your journey into football analytics. For enthusiastic fans, our book may be enough to get you up and running. For people aspiring to be quantitative football analysts, our book will hopefully serve as a springboard. For people seeking to become a data scientist or improve their data science skills, our book provides worked examples of how data can be used to answer questions. We specifically teach the following:

- How to visualize data
- How to summarize data
- How to model data
- How to present the results of data analysis
- How to use the previous techniques to tell a story

Conventions Used in This Book

The following typographical conventions are used in this book:

Italic
: Signifies new terms, URLs, email addresses, filenames, and file extensions.

`Constant width`
: Used for program listings, as well as within paragraphs to refer to program elements such as variable or function names, databases, data types, environment variables, statements, and keywords.

`Constant width bold`
: Shows commands or other text that should be typed literally by the user.

`Constant width italic`
: Shows text that should be replaced with user-supplied values or by values determined by context.

This element signifies a tip or suggestion.

This element signifies a general note.

This element indicates a warning or caution.

Using Code Examples

Supplemental material (code examples, exercises, etc.) is available for download at *https://github.com/raerickson/football_book_code*.

If you have a technical question or a problem using the code examples, please send an email to *support@oreilly.com*.

This book is here to help you get your job done. In general, if example code is offered with this book, you may use it in your programs and documentation. You do not need to contact us for permission unless you're reproducing a significant portion of the code. For example, writing a program that uses several chunks of code from this book does not require permission. Selling or distributing examples from O'Reilly books does require permission. Answering a question by citing this book and quoting example code does not require permission. Incorporating a significant amount of example code from this book into your product's documentation does require permission.

We appreciate, but generally do not require, attribution. An attribution usually includes the title, author, publisher, and ISBN. For example: "*Football Analytics with Python and R* by Eric A. Eager and Richard A. Erickson (O'Reilly). Copyright 2023 Eric A. Eager and Richard A. Erickson, 978-1-492-09962-8."

If you feel your use of code examples falls outside fair use or the permission given above, feel free to contact us at *permissions@oreilly.com*.

O'Reilly Online Learning

O'REILLY® For more than 40 years, *O'Reilly Media* has provided technology and business training, knowledge, and insight to help companies succeed.

Our unique network of experts and innovators share their knowledge and expertise through books, articles, and our online learning platform. O'Reilly's online learning platform gives you on-demand access to live training courses, in-depth learning paths, interactive coding environments, and a vast collection of text and video from O'Reilly and 200+ other publishers. For more information, visit *https://oreilly.com*.

How to Contact Us

Please address comments and questions concerning this book to the publisher:

O'Reilly Media, Inc.
1005 Gravenstein Highway North
Sebastopol, CA 95472
800-889-8969 (in the United States or Canada)
707-829-7019 (international or local)
707-829-0104 (fax)
support@oreilly.com
https://www.oreilly.com/about/contact.html

We have a web page for this book, where we list errata, examples, and any additional information. You can access this page at *https://oreil.ly/football-analytics*.

For news and information about our books and courses, visit *https://oreilly.com*.

Find us on LinkedIn: *https://linkedin.com/company/oreilly-media*

Follow us on Twitter: *https://twitter.com/oreillymedia*

Watch us on YouTube: *https://youtube.com/oreillymedia*

Acknowledgments

We thank our editors at O'Reilly, including Michelle Smith, Corbin Collins, Clare Laylock, and Aleeya Rahman for their support. We also thank Sharon Wilkey for copyediting our book, Larry Baker at Amnet Systems LLC for proofreading, and Cheryl Lenser at nSight, Inc. for creating our index. We thank Nick Adams and Danny Elfanbaum for technical assistance with the O'Reilly Atlas system. We thank Boyan Angelov, Richie Cotton, Matthew Coller, Molly Creagar, Ryan Day, Haley English, Chester Ismay, Kaelen Medeiros, George Mound, John Oliver, and Tobias Zwingmann for technical feedback. We also thank Richie Cotton for tips on writing a successful book proposal.

Eric would like to thank his wife, Stephanie, for her patience and constant support of his dreams, regardless of how crazy. He would also like to thank Neil Hornsby, whose vision in building PFF gave him a platform that has been the foundation for everything he's done until now, and Thomas Dimitroff, for answering his email back in the fall of 2020. He'd also like to thank Paul and Jack Jones for their vision in starting SumerSports in 2022. Finally, Eric would like to thank his parents, who, despite not being huge football fans themselves, kindled the flames of his passion throughout his childhood.

Figure P-3. Margo (left) and Sadie (right) walking. Who is walking whom? (Photo by Richard Erickson)

Richard thanks his daughter, Margo, for sleeping well so he could write this after she went down for the night. He also thanks Sadie for foregoing walks while he was writing, as well as patiently (and impatiently) reminding him to take stretch breaks (Figure P-3). Richard also thanks his parents for raising him with curiosity and his brother for prodding him to learn to program and for help with this book's proposal. Lastly, Richard thanks Tom Horvath and others from Hale, Skemp, Hanson, Skemp, & Sleik for support while writing this book.

CHAPTER 1
Football Analytics

American football (also known as *gridiron football* or *North American football* and henceforth simply called *football*) is undergoing a drastic shift toward the quantitative. Prior to the last half of a decade or so, most of football analytics was confined to a few seminal pieces of work. Arguably the earliest example of analytics being used in football occurred when former Brigham Young University, Chicago Bears, Cincinnati Bengals, and San Diego Chargers quarterback Virgil Carter created the notion of an *expected point* as coauthor of the 1971 paper "Technical Note: Operations Research in Football" (*https://oreil.ly/CU0I6*) before he teamed with the legendary Bill Walsh as the first quarterback to execute what is now known as the *West Coast offense*.

The idea of an *expected point* is incredibly important in football, as the game by its very nature is discrete: a collection of a finite number of plays (also called *downs*) that require the offense to go a certain distance (in yards) before having to surrender the ball to the opposing team. If the line to gain is the opponent's end zone, the offense scores a touchdown, which is worth, on average, seven points after a post-touchdown conversion. Hence, the *expected point* provides an estimated, or *expected* value for the number of *points* you would expect a team to score given the current game situation on that drive.

Football statistics have largely been confined to offensive players, and have been doled out in the currency of yards gained and touchdowns scored. The problem with this is obvious. If a player catches a pass to gain 7 yards, but 8 are required to get a first down or a touchdown, the player did not gain a first down. Conversely, if a player gains 5 yards, when 5 are required to get a first down or a touchdown, the player gained a first down. Hence, "enough" yards can be better than "more" yards depending on the context of the play. As a second example, if it takes a team two plays to travel 70 yards to score a touchdown, with one player gaining the first 65 yards and the second gaining the final 5, why should the second player get all the credit for the score?

In 1988, Bob Carroll, Pete Palmer, and John Thorn wrote *The Hidden Game of Football* (Grand Central Publishing), which further explored the notions of expected points. In 2007, Brian Burke, who was a US Navy pilot before creating the Advanced Football Analytics website (*http://www.advancedfootballanalytics.com*), formulated the expected-points and expected-points-added approach, along with building a win probability model responsible for some key insights, including the 4th Down Bot (*https://oreil.ly/Y9lEV*) at the *New York Times* website. Players may be evaluated by the number of expected points or win probability points added to their teams when those players did things like throw or catch passes, run the ball, or sack the quarterback.

The work of Burke inspired the open source work of Ron Yurko, Sam Ventura, and Max Horowitz of Carnegie Mellon University. The trio built `nflscrapR`, an R package that scraped NFL play-by-play data. The `nflscrapR` package was built to display their own versions of *expected points added* (EPA) and *win probability* (WP) models. Using this framework, they also replicated the famous *wins above replacement* (WAR) framework from baseball for quarterbacks, running backs, and wide receivers, which was published in 2018. This work was later extended using different data and methods by Eric and his collaborator George Chahouri in 2020. Eric's version of WAR, and its analogous model for college football, are used throughout the industry to this day.

The `nflscrapR` package served as a catalyst for the popularization of modern tools that use data to study football, most of which use a framework that will be replicated constantly throughout this book. The process of building an *expectation* for an outcome—in the form of points, completion percentage, rushing yards, draft-pick outcome, and many more—and measuring players or teams via the *residual* (that is, the difference between the value expected by the model and the observed value) is a process that transcends football. In soccer, for example, expected goals (xG) are the cornerstone metric upon which players and clubs are measured in the sport known as "the Beautiful Game" (*https://oreil.ly/6ol1p*). And shot quality—the expected rate at which a shot is made in basketball—is a ubiquitous measure for players and teams on the hardwood. The features that go into these models, and the forms that they take, are the subject of constant research whose surface we will scratch in this book.

The rise of tools like `nflscrapR` allowed more people to show their analytical skills and flourish in the public sphere. Analysts were hired based on their tweets on Twitter because of their ingenious approaches to measuring team performance. Decisions to punt or go for it on a fourth down were evaluated by how they affected the team's win probability. Ben Baldwin and Sebastian Carl created a spin-off R package, `nflfastR`. This package updated the models of Yurko, Ventura, and Horowitz, along with adding many of their own—models that we'll use in this book. More recently, the data contained in the `nflfastR` package has been cloned into Python via the `nfl_data_py` package by Cooper Adams.

We hope that this book will give you the basic tools to approach some of the initial problems in football analytics and will serve as a jumping-off point for future work.

 People looking for the cutting edge of sports analytics, including football, may want to check out the MIT Sloan Sports Analytics Conference. Since its founding in 2006, Sloan has emerged as a leading venue for the presentation of new tools for football (and other) analytics. Other, more accessible conferences, like the Carnegie Mellon Sports Analytics Conference and New England Statistics Symposium, are fantastic places for students and practitioners to present their work. Most of these conferences have hackathons for people looking to make an impression on the industry.

Baseball Has the Three True Outcomes: Does Football?

Baseball pioneered the use of quantitative metrics, and the creation of the Society for American Baseball Research (SABR) led to the term *sabermetrics* to describe baseball analysis. Because of this long history, we start by looking at the metrics commonly used in baseball—specifically, the *three true outcomes*. One of the reasons the game of baseball has trended toward the *three true outcomes* (walks, strikeouts, and home runs) is that they were the easiest to predict from one season to the next. What batted balls did when they were in play was noisier and was the source of much of the variance in perceived play from one year to the next. The three true outcomes of baseball have also been the source for more elaborate data-collection methods and subsequent analysis in an attempt to tease additional signals from batted-ball data.

Stability analysis is a cornerstone of team and player evaluation. *Stable* production is sticky or repeatable and is the kind of production decision-makers should want to buy into year in and year out. *Stability analysis* therefore examines whether something is stable—in our case, football observation and model outputs, and you will use this analysis in "Player-Level Stability of Passing Yards per Attempt" on page 37, "Is RYOE a Better Metric?" on page 73, "Analyzing RYOE" on page 100, and "Is CPOE More Stable Than Completion Percentage?" on page 128. On the other hand, how well a team or player does in *high-leverage situations* (plays that have greater effects on the outcome of the game, such as converting third downs) can have an outsized impact on win-loss record or eventual playoff fate, but if it doesn't help us predict what we want year in and year out, it might be better to ignore, or sell such things to other decision-makers.

Using play-by-play data from nflfastR, Chapter 2 shows you how to slice and dice football passing data into subsets that partition a player's performance into stable and unstable production. Through exploratory data analysis techniques, you can see whether any players break the mold and what to do with them. We preview how this work can aid in the process of feature engineering for prediction in Chapter 2.

Do Running Backs Matter?

For most of the history of football, the best players played running back (in fact, early football didn't include the forward pass until President Teddy Roosevelt worked with college football to introduce passing to make the game safer in 1906). The importance of the running back used to be an accepted truism across all levels of football, until the forward pass became an integral part of the game. Following the forward pass, rule and technology changes—along with Carter (mentioned earlier in this chapter) and his quarterbacks coach, Walsh—made throwing the football more efficient relative to running the football.

Many of our childhood memories from the 1990s revolve around Emmitt Smith and Barry Sanders trading the privilege of being the NFL rushing champion every other year. College football fans from the 1980s may remember Herschel Walker giving way to Bo Jackson in the Southeastern Conference (SEC). Even many younger fans from the 2000s and 2010s can still remember Adrian Peterson earning the last nonquarterback most valuable player (MVP) award. During the 2012 season, he rushed for over 2,000 yards while carrying an otherwise-bad Minnesota Vikings team to the playoffs.

However, the current prevailing wisdom among football analytics folks is that the running back position does not matter as much as other positions. This is for a few reasons. First, running the football is not as efficient as passing. This is plain to see with simple analyses using yards per play, but also through more advanced means like EPA. Even the worst passing plays usually produce, on average, more yards or expected points per play than running.

Second, differences in the actual player running the ball does not elicit the kind of change in rushing production that similar differences do for quarterbacks, wide receivers, or offensive or defensive linemen. In other words, additional resources used to pay for the services of this running back over that running back are probably not worth it, especially if those resources can be used on other positions. The marketplace that is the NFL has provided additional evidence that this is true, as we have seen running back salaries and draft capital used on the position decline to lows not previously seen.

This didn't keep the New York Giants from using the second-overall pick in the 2018 NFL Draft on Pennsylvania State University's Saquon Barkley, which was met with jeers from the analytics community, and a counter from Giants General Manager Dave Gettleman. In a post-draft press conference for the ages, Gettleman, sitting next to reams of bound paper, made fun of the analytics jabs toward his pick by mimicking a person typing furiously on a typewriter.

Chapters 3 and 4 look at techniques for controlling play-by-play rushing data for a situation to see how much of the variability in rushing success has to do with the player running the ball.

How Data Can Help Us Contextualize Passing Statistics

As we've stated previously, the passing game dominates football, and in Appendix B, we show you how to examine the basics of passing game data. In recent years, analysts have taken a deeper look into what constitutes accuracy at the quarterback position because raw completion percentage numbers, even among quarterbacks who aren't considered elite players, have skyrocketed. The work of Josh Hermsmeyer with the Baltimore Ravens and later FiveThirtyEight (*https://oreil.ly/IYs71*) established the significance of *air yards*, which is the distance traveled by the pass from the line of scrimmage to the intended receiver.

While Hermsmeyer's initial research was in the fantasy football space, it spawned a significant amount of basic research into the passing game, giving rise to metrics like *completion percentage over expected* (*CPOE*), which is one of the most predictive quarterback metrics about quarterback quality available today.

In Chapter 5, we introduce generalized linear models in the form of logistic regression. You'll use this to estimate the completion probability of a pass, given multiple situational factors that affect a throw's expected success. You'll then look at a player's *residuals* (that is, how well a player actually performs compared to the model's prediction for that performance) and see whether there is more or less stability in the residuals—the CPOE—than in actual completion percentage.

Can You Beat the Odds?

In 2018, the Professional and Amateur Sports Protection Act of 1992 (PASPA), which had banned sports betting in the United States (outside of Nevada), was overturned by the US Supreme Court. This court decision opened the floodgates for states to make legal what many people were already doing illegally: betting on football.

The difficult thing about sports betting is the *house advantage*—referred to as the *vigorish*, or *vig*—which makes it so that a bettor has to win more than 50% of their bets to break even. Thus, a cost exists for simply playing the game that needs to be overcome in order to beat the sportsbook (or simply the *book* for short).

American football is the largest gambling market in North America. Most successful sports bettors in this market use some form of analytics to overcome this house advantage. Chapter 6 examines the passing touchdowns per game prop market, which shows how a bettor would arrive at an internal price for such a market and compare it to the market price.

Do Teams Beat the Draft?

Owners, fans, and the broader NFL community evaluate coaches and general managers based on the quality of talent that they bring to their team from one year to the next. One complaint against New England Patriots Coach Bill Belichick, maybe the best nonplayer coach in the history of the NFL, is that he has not drafted well in recent seasons. Has that been a sequence of fundamental missteps or just a run of bad luck?

One argument in support of coaches such as Belichick may be "Well, they are always drafting in the back of the draft, since they are usually a good team." Luckily, one can use math to control for this and to see if we can reject the hypothesis that all front offices are equally good at drafting after accounting for draft capital used. *Draft capital* comprises the resources used during the NFL Draft—notably, the number of picks, pick rounds, and pick numbers.

In Chapter 7, we scrape publicly available draft data and test the hypothesis that all front offices are equally good at drafting after accounting for draft capital used, with surprising results. In Chapter 8, we scrape publicly available NFL Scouting Combine data and use dimension-reduction tools and clustering to see how groups of players emerge.

Tools for Football Analytics

Football analytics, and more broadly, data science, require a diverse set of tools. Successful practitioners in these fields require an understanding of these tools. Statistical programming languages, like Python and R, are a backbone of our data science toolbox. These languages allow us to clean our datasets, conduct our analyses, and readily reuse our methods.

Although many people commonly use spreadsheets (such as Microsoft Excel or Google Sheets) for data cleaning and analysis, we find spreadsheets do not scale well. For example, when working with large datasets containing tracking data, which can include thousands of rows of data per play, spreadsheets simply are not up to the task. Likewise, people commonly use business intelligence (BI) tools such as Microsoft

Power BI and Tableau because of their power and ability to scale. But these tools tend to focus on point-and-click methods and require licenses, especially for commercial use.

Programming languages also allow for easy reuse because copying and pasting formulas in spreadsheets can be tedious and error prone. Lastly, spreadsheets (and, more broadly, point-and-click tools) allow undocumented errors. For example, spreadsheets do not have a way to catch a copying and pasting mistake. Furthermore, modern data science tools allow code, data, and results to be blended together in easy-to-use interfaces. Common languages include Python, R, Julia, MATLAB, and SAS. Additional languages continue to appear as computer science advances.

As practitioners of data science, we use R and Python daily for our work, which has collectively spanned the space of applied mathematics, applied statistics, theoretical ecology and, of course, football analytics. Of the languages listed previously, Python and R offer the benefit of larger user bases (and hence likely contain the tools and models we need). Both R and Python (as well as Julia) are open source. As of this writing, Julia does not have the user base of R or Python, but it may either be the cutting edge of statistical computing, a dead end that fizzles out, or possibly both.

Open source means two types of freedom. First, anybody can access all the code in the language, like free speech. This allows volunteers to help improve the language, such as ensuring that users can debug the code and extend the language through add-on packages (like the `nflfastR` package in R or the nfl_data_py package in Python). Second, open source also offers the benefit of being free to use for users, like free drinks. Hence users do not need to pay thousands of dollars annually in licensing fees. We were initially trained in R but have learned Python over the course of our jobs. Either language is well suited for football analytics (and sports analytics in general).

 Appendix A includes instructions for obtaining R and Python for those of you who do not currently have access to these languages. This includes either downloading and installing the programs or using web-hosted resources. The appendix also describes programs to help you more easily work with these languages, such as editors and integrated development environments (IDEs).

We encourage you to pick one language for your work with this book and learn that language well. Learning a second programming language will be easier if you understand the programming concepts behind a first language. Then you can relate the concepts back to your understanding of your original computer language.

For readers who want to learn the basics of programming before proceeding with our book, we recommend Al Sweigart's *Invent Your Own Computer Games with Python*, 4th edition (No Starch Press, 2016) or Garrett Grolemund's *Hands-On Programming with R* (O'Reilly, 2014). Either resource will hold your hand to help you learn the basics of programming.

Although many people pick favorite languages and sometimes argue about which coding language is better (similar to Coke versus Pepsi or Ford versus General Motors), we have seen both R and Python used in production and also used with large data and complex models. For example, we have used R with 100 GB files on servers with sufficient memory. Both of us began our careers coding almost exclusively in R but have learned to use Python when the situation has called for it. Furthermore, the tools often have complementary roles, especially for advanced methods, and knowing both languages lets you have options for problems you may encounter.

When picking a language, we suggest you *use what your friends use*. If all your friends speak Spanish, communicating with them if you learn Spanish will probably be easier as well. You can then teach them your native language too. Likewise, the same holds for programming: your friends can then help you debug and troubleshoot. If you still need help deciding, open up both languages and play around for a little bit. See which one you like better. Personally, we like R when working with data, because of R's data manipulation tools, and Python when building and deploying new models because of Python's cleaner syntax for writing functions.

First Steps in Python and R

If you are familiar with R and Python, you'll still benefit from skimming this section to see how we teach a tool you are familiar with.

Opening a computer terminal may be intimidating for many people. For example, many of our friends and family will walk by our computers, see code up on the screens, and immediately turn their heads in disgust (Richard's dad) or fear (most other people). However, terminals are quite powerful and allow more to be done with less, once you learn the language. This section will help you get started using Python or R.

The first step for using R or Python is either to install it on your computer or use a web-based version of the program. Various options exist for installing or otherwise accessing Python and R and then using them on your computer. Appendix A contains steps for this as well as installation options.

People, like Richard, who follow the Green Bay Packers are commonly called *Cheeseheads*. Likewise, people who use Python are commonly called *Pythonistas*, and people who use R are commonly called *useRs*.

Once you have you access to R or Python, you have an expensive graphing calculator (for example, your $1,000 laptop). In fact, both Eric and Richard, in lieu of using an actual calculator, will often calculate silly things like point spreads or totals in the console if in need of a quick calculation. Let's see some things you can do. Type **2 + 2** in either the Python or R console:

```
2 + 2
```

Which results in:

```
4
```

People use comments to leave notes to themselves and others in code. Both Python and R use the # symbol for comments (the *pound symbol* for the authors or *hashtag* for younger readers). Comments are text (within code) that the computer does not read but that help humans to understand the code. In this book, we will use two comment symbols to tell you that a code block is Python (## Python) or R (## R)

You may also save numbers as variables. In Python, you could define z to be 2 and then reuse z and divide by 3:

```
## Python
z = 2
z / 3
```

Resulting in:

```
0.6666666666666666
```

In R, either <- or = may be used to create variables. We use <- for two reasons. First, in this book this helps you see the difference between R and Python code. Second, we use this style in our day-to-day programming as well. Chapter 9 discusses code styles more. Regardless of which operator you use, be consistent with your programming style in any language. Your future self (and others who read your code) will thank you.

In R, you can also define z to be 2 and then reuse z and divide by 3:

```
## R
z <- 2
z / 3
```

Resulting in:

```
[1] 0.6666667
```

Python and R format outputs differently. Python does not round up and includes more digits. Conversely, R shows fewer digits and rounds up.

Example Data: Who Throws Deep?

Now that you have seen some basics in R, let's dive into an example with football data. You will use the nflfastR data for many of the examples in this book. This data may be installed as an R package or as the Python package nfl_data_py. Specifically, we will explore the broad (and overly simple) question "Who were the most aggressive quarterbacks in 2021?" We will start off introducing the package using R because the data originated with R.

Both Python and R have flourished because they readily allow add-on packages. Conda exists as one tool for managing these add-ons. Chapter 9 and Appendix A discuss these add-ons in greater detail. In general, you can install packages in Python by typing **pip install** *package name* or **conda install** *package name* in the terminal (such as the bash shell on Linux, Zsh shell on macOS, or command prompt on Microsoft Windows). Sometimes you will need to use pip3, depending on your operating system's configuration, if you are using the pip package manager system. For a concrete example, to install the seaborn package, you could type **pip install seaborn** in your terminal. In general, packages in R can be installed by opening R and then typing **install.packages("***package name***")**. For example, to install the tidyverse collection of packages, open R and run **install.pack ages("tidyverse")**.

nflfastR in R

Starting with R, install the nflfastR package:

```
## R
install.packages("nflfastR")
```

Using single quotation marks around a name, such as 'x', or double quotes, such as "x", are both acceptable to languages such as Python or R. Make sure the opening and closing quotes match. For example, 'x" would not be acceptable. You may use both single and double quotes to place quotes inside of quotes. For example, in a figure caption, you might write, "Panthers' points earned" or 'Air temperature ("true temperature")'. Or in Python, you can use a combination of quotes later for inputs such as "team == 'GB'" because you'll need to nest quotes inside of quotes.

Next, load this package as well as the tidyverse, which gives you tools to manipulate and plot the data:

```
## R
library("tidyverse")
library("nflfastR")
```

Base R contains dataframes as `data.frame()`. We use tibbles from the tidyverse instead, because these print nicer to screens and include other useful features. Many users consider base R's `data.frame()` to be a legacy object, although you will likely see these objects when looking at help files and examples on the web. Lastly, you might see the `data.table` package in R. The `data.table` extension of dataframes is similar to a tibble and works better with larger data (for example, 10 GB or 100 GB files) and has a more compact coding syntax, but it comes with the trade-off of being less user-friendly compared to tibbles. In our own work, we use a data.table rather than a tibble or data.frame when we need high performance at the trade-off of code readability.

Once you've loaded the packages, you need to load the data from each play, or the *play-by-play* (pbp) data, for the 2021 season. Use the `load_pbp()` function from `nflfastR` and call the data `pbp_r` (the _r ending helps you tell that the code is from an R example in this book):

```
## R
pbp_r <- load_pbp(2021)
```

 We generally include _py in the name of Python dataframes and _r in the names of R dataframes to help you identify the language for various code objects.

After loading the data as `pbp_r`, pass (or *pipe*) the data along to be *filtered* by using `|>`. Use the `filter()` function to select only data where passing plays occurred (`play_type == "pass"`) and where `air_yards` are not missing, or `NA` in R syntax (in plain English, the pass had a recorded depth). Chapter 2, Appendix B, and Appendix C cover data manipulation more, and most examples in this book use data wrangling to format data. So right now, simply type this code. You can probably figure out what the code is doing, but don't worry about understanding it too much:

```
## R
pbp_r_p <-
    pbp_r |>
    filter(play_type == 'pass' & !is.na(air_yards))
```

Now you'll look at the average depth of target (aDOT), or mean air yards per pass, for every quarterback in the NFL in 2021 who threw 100 or more passes with a designated depth. To avoid multiple players who have the same name, which happens more than you'd think, you'll summarize by both player ID and player name.

First, *group by* both the `passer_id` and `passer`. Then *summarize* to calculate the number of plays (`n()`) and mean air yards per pass (`adot`) per player. Also, *filter* to include only players with 100 or more plays and to remove any rows without a passer name (specifically, those with missing or `NA` values).

With this and the previous example commands, the function `is.na(passer)` checks whether value in the `passer` column has the value `NA` and returns `TRUE` for columns with an `NA` value. Appendix B covers this logic and terminology in greater detail. Next, an exclamation point (`!`) turns this expression into the opposite of *not missing value*, so that you keep cells with a value. As an aside, we, the authors, find the use of double negatives confusing as well. Lastly, arrange by the `adot` values and then print all (or infinity, `Inf`) values:

```
## R
pbp_r_p |>
    group_by(passer_id, passer) |>
    summarize(n = n(), adot = mean(air_yards)) |>
    filter(n >= 100 & !is.na(passer)) |>
    arrange(-adot) |>
    print(n = Inf)
```

Resulting in:

```
[Entire table]

A tibble: 42 × 4
# Groups:   passer_id [42]
   passer_id  passer              n  adot
   <chr>      <chr>          <int> <dbl>
 1 00-0035704 D.Lock           110 10.2
 2 00-0029263 R.Wilson         400  9.89
 3 00-0036945 J.Fields         268  9.84
 4 00-0034796 L.Jackson        378  9.34
 5 00-0036389 J.Hurts          473  9.19
 6 00-0034855 B.Mayfield       416  8.78
 7 00-0026498 M.Stafford       740  8.51
 8 00-0031503 J.Winston        161  8.32
 9 00-0029604 K.Cousins        556  8.23
10 00-0034857 J.Allen          708  8.22
11 00-0031280 D.Carr           676  8.13
12 00-0031237 T.Bridgewater    426  8.04
13 00-0019596 T.Brady          808  7.94
14 00-0035228 K.Murray         515  7.94
15 00-0036971 T.Lawrence       598  7.91
16 00-0036972 M.Jones          557  7.90
17 00-0033077 D.Prescott       638  7.81
18 00-0036442 J.Burrow         659  7.75
19 00-0023459 A.Rodgers        556  7.73
20 00-0031800 T.Heinicke       491  7.69
21 00-0035993 T.Huntley        185  7.68
22 00-0032950 C.Wentz          516  7.64
```

```
23 00-0029701 R.Tannehill            554  7.61
24 00-0037013 Z.Wilson               382  7.57
25 00-0036355 J.Herbert              671  7.55
26 00-0033119 J.Brissett             224  7.55
27 00-0033357 T.Hill                 132  7.44
28 00-0028118 T.Taylor               149  7.43
29 00-0030520 M.Glennon              164  7.38
30 00-0035710 D.Jones                360  7.34
31 00-0036898 D.Mills                392  7.32
32 00-0031345 J.Garoppolo            511  7.31
33 00-0034869 S.Darnold              405  7.26
34 00-0026143 M.Ryan                 559  7.16
35 00-0032156 T.Siemian              187  7.13
36 00-0036212 T.Tagovailoa           387  7.10
37 00-0033873 P.Mahomes              780  7.08
38 00-0027973 A.Dalton               235  6.99
39 00-0027939 C.Newton               126  6.97
40 00-0022924 B.Roethlisberger       647  6.76
41 00-0033106 J.Goff                 489  6.44
42 00-0034401 M.White                132  5.89
```

The `adot` value, a commonly used measure of quarterback aggressiveness, gives a quantitative approach to rank quarterbacks by their aggression, as measured by mean air yards per pass (can you think of other ways to measure aggressiveness that pass depth alone leaves out?). Look at the results and think, do they make sense to you, or are you surprised, given your personal opinions of quarterbacks?

 If you get unexpected errors on any of the commands, double-check that you are in the correct language environment. You may be trying to use Python in the R environment or R in the Python environment.

nfl_data_py in Python

In Python, the `nfl_data_py` package by Cooper Adams exists as a clone of the R `nflfastR` package for data. To use the data from this package, first import the `pandas` package with the alias (or short nickname) `pd` for working with data and import the `nfl_data_py` package as `nfl`:

```Python
## Python
import pandas as pd
import nfl_data_py as nfl
```

Next, tell Python to import the data for 2021 (Chapter 2 shows how to import multiple years). Note that you need to include the year in a Python list as [2021]:

```Python
## Python
pbp_py = nfl.import_pbp_data([2021])
```

As with the R code, filter the data in Python (pandas calls filtering a `query`). Python allows you to readily pass the filter criteria (`filter_crit`) into `query()` as an object, and we have you do this to save space line space. Then *group by* `passer_id` and `passer` before *aggregating* the data by using a Python dictionary (`dict()`, or `{}` for short) with the `.agg()` function:

```Python
## Python
filter_crit = 'play_type == "pass" & air_yards.notnull()'

pbp_py_p = (
    pbp_py.query(filter_crit)
    .groupby(["passer_id", "passer"])
    .agg({"air_yards": ["count", "mean"]})
)
```

The pandas package also requires reformatting the column heads via a `list()` function and changing the header from being two rows to a single row via map(). Next, print the outputs after sorting by the mean of the air yards via the `query()` function (`to_string()` allows all the outputs to be printed):

```Python
## Python
pbp_py_p.columns = list(map("_".join, pbp_py_p.columns.values))
sort_crit = "air_yards_count > 100"
print(
    pbp_py_p.query(sort_crit)\
    .sort_values(by="air_yards_mean", ascending=[False])\
    .to_string()
)
```

This results in:

```
                     air_yards_count  air_yards_mean
passer_id  passer
00-0035704 D.Lock              110       10.154545
00-0029263 R.Wilson           400        9.887500
00-0036945 J.Fields           268        9.835821
00-0034796 L.Jackson          378        9.341270
00-0036389 J.Hurts            473        9.190275
00-0034855 B.Mayfield         416        8.776442
00-0026498 M.Stafford         740        8.508108
00-0031503 J.Winston          161        8.322981
00-0029604 K.Cousins          556        8.228417
00-0034857 J.Allen            708        8.224576
00-0031280 D.Carr             676        8.128698
00-0031237 T.Bridgewater      426        8.037559
00-0019596 T.Brady            808        7.941832
00-0035228 K.Murray           515        7.941748
00-0036971 T.Lawrence         598        7.913043
00-0036972 M.Jones            557        7.901257
00-0033077 D.Prescott         638        7.811912
00-0036442 J.Burrow           659        7.745068
00-0023459 A.Rodgers          556        7.730216
```

00-0031800	T.Heinicke	491	7.692464
00-0035993	T.Huntley	185	7.675676
00-0032950	C.Wentz	516	7.641473
00-0029701	R.Tannehill	554	7.606498
00-0037013	Z.Wilson	382	7.565445
00-0036355	J.Herbert	671	7.554396
00-0033119	J.Brissett	224	7.549107
00-0033357	T.Hill	132	7.439394
00-0028118	T.Taylor	149	7.429530
00-0030520	M.Glennon	164	7.378049
00-0035710	D.Jones	360	7.344444
00-0036898	D.Mills	392	7.318878
00-0031345	J.Garoppolo	511	7.305284
00-0034869	S.Darnold	405	7.259259
00-0026143	M.Ryan	559	7.159213
00-0032156	T.Siemian	187	7.133690
00-0036212	T.Tagovailoa	387	7.103359
00-0033873	P.Mahomes	780	7.075641
00-0027973	A.Dalton	235	6.987234
00-0027939	C.Newton	126	6.968254
00-0022924	B.Roethlisberger	647	6.761978
00-0033106	J.Goff	489	6.441718
00-0034401	M.White	132	5.886364

Hopefully, this chapter whet your appetite for using math to examine football data. We glossed over some of the many topics you will learn about in future chapters such as data sorting, summarizing data, and cleaning data. You have also had a chance to compare Python and R for basic tasks for working with data, including modeling. Appendix B also dives deeper into the air-yards data to cover basic statistics and data wrangling.

Data Science Tools Used in This Chapter

This chapter covered the following topics:

- Obtaining data from one season by using the `nflfastR` package either directly in R or via the `nfl_data_py` package in Python
- Using `filter()` in R or `query()` in Python to select and create a subset of data for analysis
- Using `summarize()` to group data in R with the help of `group_by()`, and aggregating (`agg()`) data by groups in Python with the help of `groupby()`
- Printing dataframe outputs to your screen to help you look at data
- Removing missing data by using `is.na()` in R or `notnull()` in Python

Suggested Readings

If you get really interested in analytics without the programming, here are some sources we read to develop our philosophy and strategies for football analytics:

- *The Hidden Game of Football: A Revolutionary Approach to the Game and Its Statistics* by Bob Carroll et al. (University of Chicago Press, 2023). Originally published in 1988, this cult classic introduces the numerous ideas that were later formulated into the cornerstone of what has become modern football analytics.

- *Moneyball: The Art of Winning an Unfair Game* by Michael Lewis (W.W. Norton & Company, 2003). Lewis describes the rise of analytics in baseball and shows how the stage was set for other sports. The book helps us think about how modeling and data can help guide sports. A movie was made of this book as well.

- *The Signal and the Noise: Why So Many Predictions Fail, but Some Don't* by Nate Silver (Penguin, 2012). Silver describes why models work in some instances and fail in others. He draws upon his experience with poker, baseball analytics, and running the political prediction website FiveThirtyEight. The book does a good job of showing how to think quantitatively for big-picture problems without getting bogged down by the details.

Lastly, we encourage you to read the documentation for the `nflfastR` package (*https://www.nflfastr.com*). Diving into this package will help you better understand much of the data used in this book.

## CHAPTER 2

Exploratory Data Analysis: Stable Versus Unstable Quarterback Statistics

In any field of study, a level of intuition (commonly known as a *gut feeling*) can exist that separates the truly great subject-matter experts from the average ones, the average ones from the early-career professionals, or the early-career professionals from the novices. In football, that skill is said to manifest itself in player evaluation, as some scouts are perceived to have a knack for identifying talent through great intuition earned over years of honing their craft. Player traits that translate from one situation to the next—whether from college football to the professional ranks, or from one coach's scheme to another's—require recognition and further investigation, while player outcomes that cannot be measured (at least using current data and tools) are discarded. Experts in player evaluation also know how to properly communicate the fruits of their labor in order to gain maximum collective benefit from it.

While traditional scouting and football analytics are often considered at odds with each other, the statistical evaluation of players requires essentially the same process. Great football analysts are able to, when evaluating a player's data (or multiple players' data), find the right data specs to interrogate, production metrics to use, situational factors to control for, and information to discard. How do you acquire such an acumen? The same way a scout does. Through years of deliberate practice and refinement, an analyst gains not only a set of tools for player, team, scheme, and game evaluation but also a knack for the right question to ask at the right time.

Famous statistician John Tukey noted (*https://oreil.ly/RVyMp*), "Far better an approximate answer to the right question, which is often vague, than an exact answer to the wrong question, which can always be made precise." Practically, this quote illustrates that in football, or broader data science, asking the question that meets your needs is more important than using your existing data and models precisely.

One advantage that statistical approaches have over traditional methods is that they are scalable. Once an analyst develops a tried-and-true method for player evaluation, they can use the power of computing to run that analysis on many players at once—a task that is incredibly cumbersome for traditional scouting methods.

In this chapter, we give you some of the first tools necessary to develop a knack for football evaluation using statistics. The first idea to explore in this topic is *stability*. Stability is important when evaluating anything, especially in sports. This is because stability measures provide a comparative way to determine how much of a skill is *fundamental* to the player—how much of what happened in a given setting is transferable to another setting—and how much of past performance can be attributed to *variance*. In the words of FiveThirtyEight founder Nate Silver, stability analysis helps us tease apart what is *signal* and what is *noise*. If a player does very well in the stable components of football but poorly in the unstable ones, they might be a *buy-low* candidate—a player who is underrated in the marketplace. The opposite, a player who performs well in the unstable metrics but poorly in the stable ones, might be a *sell-high* player.

Exact definitions of stability vary based on specific contexts, but in this book we refer to the stability of an evaluation metric as the metric's consistency over a predetermined time frame. For example, for fantasy football analysts, that time frame might be week to week, while for an analyst building a draft model for a team, it might be from a player's final few seasons in college to his first few seasons as a pro. The football industry generally uses Pearson's correlation coefficient, or its square, the coefficient of determination, to measure stability. A *Pearson's correlation coefficient* ranges from –1 to 1. A value of 0 corresponds to no correlation, whereas a value of 1 corresponds to a perfect positive correlation (two variables increase in value together), and a value of –1 corresponds to a perfect negative correlation (one variable increases as the second variable decreases).

While the precise numerical threshold for a metric to be considered stable is usually context- or era-specific, a higher coefficient means a more stable statistic. For example, pass-rushing statistics are generally pretty stable, while coverage statistics are not; Eric talked about this more in a recent paper from the Sloan Sports Analytics Conference (*https://oreil.ly/_GOBP*). If comparing two pass-rushing metrics, the less-stable one might have a lower correlation coefficient than the more-stable coverage metric.

Fantasy fans will know *stability analysis* by another term: *sticky stats*. This term arises because some statistical estimates "stick" around and are consistent through time.

Stability analysis is part of a subset of statistical analysis called *exploratory data analysis* (*EDA*), which was coined by the American statistician John Tukey. In contrast to formal modeling and hypothesis testing, EDA is an approach of analyzing datasets to summarize their main characteristics, often using statistical graphics and other data visualization methods. EDA is an often-overlooked step in the process of using statistical analysis to understand the game of football—both by newcomers and veterans of the discipline—but for different reasons.

 John Tukey also coined other terms, some that you may know or will hopefully know by the end of this book, including *boxplot* (a type of graph), *analysis of variance* (ANOVA for short; a type of statistical test), *software* (computer programs), and *bit* (the smallest unit of computer data, usually represented as 0/1; you're probably more familiar with larger units such as the byte, which is 8 bits, and larger units like gigabytes). Tukey and his students also helped the Princeton University football team implement data analysis using basic statistical methods by examining over 20 years of football data. However, the lack of modern computers limited his work, and many of the tools you learn in this book are more advanced than the methods he had access to. For example, one of his former students, Gregg Lange (*https://oreil.ly/L3MCF*), remembered how a simple mistake required him to reload 100 pounds of data cards into a computer. To read more about Tukey's life and contributions, check out "John W. Tukey: His Life and Professional Contributions" by David Brillinger in the *Annals of Statistics* (*https://oreil.ly/7Wb0a*).

Defining Questions

Asking the right questions is as important as solving them. In fact, as the Tukey quote highlights, the right answer to the wrong question is useless in and of itself, while the right question can lead you to prosperous outcomes even if you fall short of the correct answer. Learning to ask the right question is a process honed by learning from asking the wrong questions. Positive results are the spoils earned from fighting through countless negative results.

To be scientific, a question needs to be about a hypothesis that is both testable and falsifiable. For example, "Throwing deep passes is more valuable than short passes, but it's difficult to say whether or not a quarterback is good at deep passes" is a reasonable hypothesis, but to make it scientific, you need to define what "valuable" means and what you mean when we say a player is "good" (or "bad") at deep passes. To that aim, you need data.

Obtaining and Filtering Data

To study the stability of passing data, use the `nflfastR` package in R or the `nfl_data_py` package in Python. Start by loading the data from 2016 to 2022 as play-by-play, or `pbp`, using the tools you learned in Chapter 1.

Using 2016 is largely an arbitrary choice. In this case, it's the last year with a material rule change (moving the kickoff touchback up to the 25-yard line) that affected game play. Other seasons are natural breaking points, like 2002 (the last time the league expanded), 2011 (the last influential change to the league's collective bargaining agreement), 2015 (when the league moved the extra point back to the 15-yard line), 2020 (the COVID-19 pandemic, and also when the league expanded the playoffs), and 2021 (when the league moved from 16 to 17 regular-season games).

In Python, load the `pandas` and `numpy` packages as well as the `nfl_data_py` package:

```
## Python
import pandas as pd
import numpy as np
import nfl_data_py as nfl
```

 Python starts numbering at 0. R starts numbering at 1. Many an aspiring data scientist has been tripped up by this if using both languages. Because of this, you need to add **+ 1** to the input of the last value in `range()` in this example.

Next, tell Python what years to load by using `range()`. Then import the NFL data for those seasons:

```
## Python
seasons = range(2016, 2022 + 1)
pbp_py = nfl.import_pbp_data(seasons)
```

In R, first load the required packages. The `tidyverse` collection of packages helps you wrangle and plot the data. The `nflfastR` package provides you with the data. The `ggthemes` package assists with plotting formatting:

```
## R
library("tidyverse")
library("nflfastR")
library("ggthemes")
```

In R, you may use the shortcut `2016:2022` to specify the range 2016 to 2022:

```
## R
pbp_r <- load_pbp(2016:2022)
```

 With any dataset, understand the *metadata*, or the data about the data. For example, what do 0 and 1 mean? Which is yes and which is no? Or do the authors use 1 and 2 for levels? We have heard about scientific studies being retracted because the data analysts and scientists misunderstood the metadata and the uses of 1 and 2 versus the standard 0 and 1. Thus, scientists had to tell people their study was flawed because they did not understand their own data structure. For example, a 2021 article in *Significance* (*https://oreil.ly/ 9kORC*) describes an occurrence of this mistake.

To get the subset of data you need for this analysis, filter down to just the passing plays, which can be done with the following code:

```
## Python
pbp_py_p = \
    pbp_py\
    .query("play_type == 'pass' & air_yards.notnull()")\
    .reset_index()
```

In R, `filter()` the data by using the same criteria:

```
## R
pbp_r_p <-
    pbp_r |>
    filter(play_type == "pass" & !is.na(air_yards))
```

Here, `play_type` being equal to `pass` will eliminate both running plays and plays that are negated because of a penalty. Sometimes you want to include plays that have a penalty (for example, if you are using a grade-based system like the one at PFF). Grade-based systems attempt to measure how well the player performed on a play independent of the final statistics of the play, so keeping data where `play_type ==` `no_play` might have value.

For the sake of this exercise, though, we have you omit such plays. You also omit plays where air_yards is NA (in R) or NULL (in Python). These plays occur when a pass is not aimed at an intended receiver because it's batted down at the line of scrimmage, thrown away, or spiked. While those passes certainly count toward a passer's final statistics, and are fundamental to who he is as a player, they are not necessarily relevant to the question being asking here.

Next, you need to do some data cleaning and wrangling.

First, define a *long* pass as a pass that has air yards greater than or equal to 20 yards, and a *short* pass as one with air yards less than 20 yards. The NFL has a categorical variable for pass length (pass_length) in data, but the classifications are not completely clear to the observer (see the exercises at the end of the chapter). Luckily, you can easily calculate this on your own (and use a different criterion if desired, such as 15 yards or 25 yards).

Second, the passing yards for incomplete passes are recorded as NA in R, or NULL in Python, but should be set to 0 for this analysis (as long as you've filtered properly previously).

In Python, the numpy (imported as np) package's where() function helps with this change. First, create the filtering criteria:

```Python
## Python
pbp_py_p["pass_length_air_yards"] = np.where(
    pbp_py_p["air_yards"] >= 20, "long", "short"
)
```

Then use the filtering criteria to replace missing values:

```Python
## Python
pbp_py_p["passing_yards"] = \
    np.where(
        pbp_py_p["passing_yards"].isnull(), 0, pbp_py_p["passing_yards"]
        )
```

In R, the ifelse() function inside mutate() allows the same change:

```R
## R
pbp_r_p <-
    pbp_r_p |>
    mutate(
        pass_length_air_yards = ifelse(air_yards >= 20, "long", "short"),
        passing_yards = ifelse(is.na(passing_yards), 0, passing_yards)
    )
```

Appendix B covers data manipulation topics such as filtering in great detail. Refer to this source if you need help better understanding our data wrangling. We are glossing over these details to help you get into the data right away with interesting questions.

Naming objects can be surprisingly hard when programming. Try to balance simple names that are easier to type with longer, more informative names. This can be especially important if you start writing scripts with longer names. The most important part of naming is to create understandable names for both others and your future self.

Summarizing Data

Briefly examine some basic numbers used to describe the passing_yards data. In Python, select the passing_yards column and then use the describe() function:

```
## Python
pbp_py_p["passing_yards"]\
    .describe()
```

Resulting in:

```
count    131606.000000
mean          7.192111
std           9.667021
min         -20.000000
25%           0.000000
50%           5.000000
75%          11.000000
max          98.000000
Name: passing_yards, dtype: float64
```

In R, take the dataframe and select (or pull()) the passing_yards column and then calculate the summary() statistics:

```
## R
pbp_r_p |>
    pull(passing_yards) |>
    summary()
```

This results in:

```
   Min. 1st Qu.  Median    Mean 3rd Qu.    Max.
-20.000   0.000   5.000   7.192  11.000  98.000
```

In the outputs, here are what the names describe (Appendix B shows how to calculate these values):

- The count (only in Python) is the number of records in the data.
- The mean in Python (Mean in R) is the arithmetic average.
- The std (only in Python) is the standard deviation.
- The min in Python or Min. in R is the lowest or minimum value.

- The 25% in Python or `1st Qu.` in R is the first-quartile value, for which one-fourth of all values are smaller.

- The `Median` (in R) or 50% (in Python) is the middle value, for which half of the values are bigger and half are smaller.

- The 75% in Python or `3rd Qu.` in R is the third-quartile value, for which three-quarters of all values are smaller.

- The `max` in Python or `Max.` in R is the largest or maximum value.

What you really want to see is a summary of the data under different values of `pass_length_air_yards`. For short passes, filter out the long passes and then summarize, in Python:

```
## Python
pbp_py_p\
    .query('pass_length_air_yards == "short"')["passing_yards"]\
    .describe()
```

Resulting in:

```
count    116087.000000
mean          6.526812
std           7.697057
min         -20.000000
25%           0.000000
50%           5.000000
75%          10.000000
max          95.000000
Name: passing_yards, dtype: float64
```

And in R:

```
## R
pbp_r_p |>
    filter(pass_length_air_yards == "short") |>
    pull(passing_yards) |>
    summary()
```

Which results in:

```
   Min. 1st Qu.  Median    Mean 3rd Qu.    Max.
-20.000   0.000   5.000   6.527  10.000  95.000
```

Likewise, you can filter to select long passes in Python:

```
## Python
pbp_py_p\
    .query('pass_length_air_yards == "long"')["passing_yards"]\
    .describe()
```

Resulting in:

```
count    15519.000000
mean        12.168761
std         17.923951
min          0.000000
25%          0.000000
50%          0.000000
75%         26.000000
max         98.000000
Name: passing_yards, dtype: float64
```

And in R:

```
## R
pbp_r_p |>
    filter(pass_length_air_yards == "long") |>
    pull(passing_yards) |>
    summary()
```

Resulting in:

```
  Min. 1st Qu.  Median    Mean 3rd Qu.    Max.
  0.00    0.00    0.00   12.17   26.00   98.00
```

The point to notice here is that the *interquartile range*, the difference between the first and third quartile, is much larger for longer passes than for short passes, even though the maximum passing yards are about the same. The minimum values are going to be higher for long passes, since it's almost impossible to gain negative yards on a pass that travels 20 or more yards in the air.

You can perform the same analysis for expected points added (EPA), which was introduced in Chapter 1. Recall that EPA is a more continuous measure of play success that uses situational factors to assign a point value to each play. You can do this in Python:

```
## Python
pbp_py_p\
    .query('pass_length_air_yards == "short"')["epa"]\
    .describe()
```

Resulting in:

```
count    116086.000000
mean          0.119606
std           1.426238
min         -13.031219
25%          -0.606135
50%          -0.002100
75%           0.959107
max           8.241420
Name: epa, dtype: float64
```

And in R:

```R
## R
pbp_r_p |>
    filter(pass_length_air_yards == "short") |>
    pull(epa) |>
    summary()
```

Which results in:

```
    Min.   1st Qu.   Median     Mean   3rd Qu.     Max.     NA's
-13.0312   -0.6061  -0.0021   0.1196   0.9591   8.2414        1
```

Likewise, you can do this for long passes in Python:

```Python
## Python
pbp_py_p\
    .query('pass_length_air_yards == "long"')["epa"]\
    .describe()
```

Resulting in:

```
count     15519.000000
mean          0.382649
std           2.185551
min         -10.477922
25%          -0.827421
50%          -0.465344
75%           2.136431
max           8.789743
Name: epa, dtype: float64
```

Or in R:

```R
## R
pbp_r_p |>
    filter(pass_length_air_yards == "long") |>
    pull(epa) |>
    summary()
```

Resulting in:

```
    Min.   1st Qu.   Median     Mean   3rd Qu.     Max.
-10.4779   -0.8274  -0.4653   0.3826   2.1364   8.7897
```

You get the same dynamic here: wider outcomes for longer passes than shorter ones. Longer passes are more *variable* than shorter passes.

Furthermore, if you look at the mean passing yards per attempt (YPA) and EPA per attempt for longer passes, they are both higher than those for short passes (while the relationship flips for the median, why is that?). Thus, on average, you can informally confirm the first part of our guiding hypothesis for the chapter: "Throwing deep passes is more valuable than short passes, but it's difficult to say whether or not a quarterback is good at deep passes."

Line breaks and white space are important for coding. These breaks help make your code easier to read. Python and R also handle line breaks differently, but sometimes both languages treat line breaks as special commands. In both languages, you often split function inputs to create shorter lines that are easier to read. For example, you can space a function as follows to break up line names and make your code easier to read:

```
## Python or R
my_plot(data=big_name_data_frame,
        x="long_x_name",
        y="long_y_name")
```

In R, make sure the comma stays on a previous line. In Python, you may need to use a \ for line breaks.

```
## Python
x =\
    2 + 4
```

Or put the entire command in parentheses:

```
## Python
x = (
    2 + 4
    )
```

You can also write one Python function per line:

```
## Python
my_out = \
    my_long_long_long_data\
    .function_1()\
    .function_2()
```

Plotting Data

While numerical summaries of data are useful, and many people are more algebraic thinkers than they are geometric ones (Eric is this way), many people need to visualize something other than numbers. Reasons we like to plot data include the following:

- Checking to make sure the data looks OK. For example, are any values too large? Too small? Do other wonky data points exist?

- Are there outliers in the data? Do they arise naturally (e.g., Patrick Mahomes in almost every passing efficiency chart) or unnaturally (e.g., a probability below 0 or greater than 1)?

- Do any broad trends emerge at first glance?

Histograms

Histograms, a type of plot, allow you to see data by summing the counts of data points into bars. These bars are called *bins*.

 If you have previous versions of the packages used in this book installed, you may need to upgrade if our code examples will not work. Conversely, future versions of packages used in this book may update how functions work. The book's GitHub page (github.com/raerickson/football_book_code) may have updated code.

In Python, we use the `seaborn` package for most plotting in the book. First, import `seaborn` by using the alias `sns`. Then use the `displot()` function to create the plot shown in Figure 2-1:

```python
## Python
import seaborn as sns
import matplotlib.pyplot as plt

sns.displot(data=pbp_py, x="passing_yards");
plt.show();
```

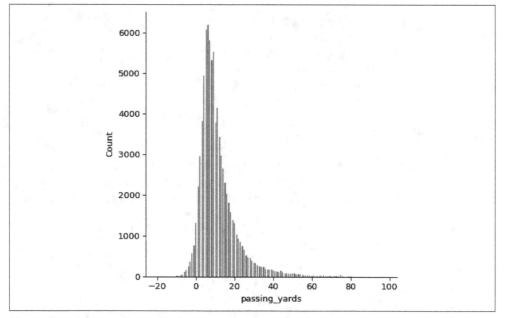

Figure 2-1. A histogram in Python using seaborn for the passing_yards variable

On macOS, you also need to include `import matplotlib.pyplot as plt` when you load other packages. Likewise, macOS users also need to include `plt.show()` for their plot to appear after their plotting code. We also found we needed to use `plt.show()` with some editors on Linux (such as Microsoft Visual Studio Code) but not others (such as JupyterLab). If in doubt, include this optional code. Running `plt.show()` will do no harm but might be needed to make your figures appear. Windows may or may not require this.

Likewise, R allows for histograms to be easily created.

Although base R comes with its own plotting tools, we use `ggplot2` for this book. The `ggplot2` tool has its own language, based on *The Grammar of Graphics* by Leland Wilkinson (Springer, 2005) and implemented in R by Hadley Wickham during his doctoral studies at Iowa State University. Pedagogically, we agree with David Robinson, who describes his reasons for teaching plotting with `ggplot2` over base R in a blog post titled "Don't Teach Built-in Plotting to Beginners (Teach ggplot2)" (*https://oreil.ly/QDtpo*).

In R, create the histogram shown in Figure 2-2 by using `ggplot2` in R with the `ggplot()` function. In the function, use the `pbp_r_p` dataset and set the aesthetic for x to be `passing_yards`. Then add the geometry `geom_histogram()`:

```
## R
ggplot(pbp_r, aes(x = passing_yards)) +
    geom_histogram()
```

Resulting in:

```
`stat_bin()` using `bins = 30`. Pick better value with `binwidth`.

Warning: Removed 257229 rows containing non-finite values (`stat_bin()`).
```

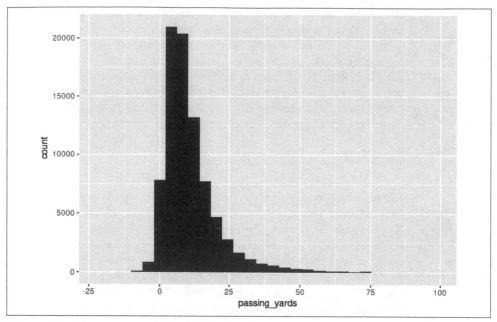

Figure 2-2. A histogram in R using ggplot2 *for the* passing_yards *variable*

 Intentionally using the wrong number of bins to hide important attributes of your data is considered fraud by the larger statistical community. Be thoughtful and intentional when you select the number of bins for a histogram. This process requires many iterations as you explore various numbers of histogram bins.

Figures 2-1 and 2-2 let you understand the basis of our data. Passing yards gained ranges from about −10 yards to about 75 yards, with most plays gaining between 0 (often an incompletion) and 10 yards. Notice that R warns you to be careful with the binwidth and the number of bins and also warns you about the removal of missing values. Rather than using the default, set each bin to be 1 yard wide. You can either ignore the second warning about missing values or filter out missing values prior to plotting to avoid the warning. With such a bin width, the data no longer looks normal, because of the many, many incomplete passes.

Next, you will make a histogram for each pass_depth_air_yards value. We will show you how to create the short pass in Python (Figure 2-3) and the long pass in R (Figure 2-4).

In Python, change the theme to be colorblind for the palette option and use a whitegrid option to create plots similar to ggplot2's black-and-white theme:

```Python
## Python
import seaborn as sns
import matplotlib.pyplot as plt

sns.set_theme(style="whitegrid", palette="colorblind")
```

Next, filter out the short passes:

```Python
## Python
pbp_py_p_short = \
    pbp_py_p\
    .query('pass_length_air_yards == "short"')
```

Then create a histogram and use `set_axis_labels` to change the plot's labels, making it look better, as shown in Figure 2-3:

```Python
## Python
# Plot, change labels, and then show the output
pbp_py_hist_short = \
    sns.displot(data=pbp_py_p_short,
                binwidth=1,
                x="passing_yards");
pbp_py_hist_short\
    .set_axis_labels(
        "Yards gained (or lost) during a passing play", "Count"
        );
plt.show();
```

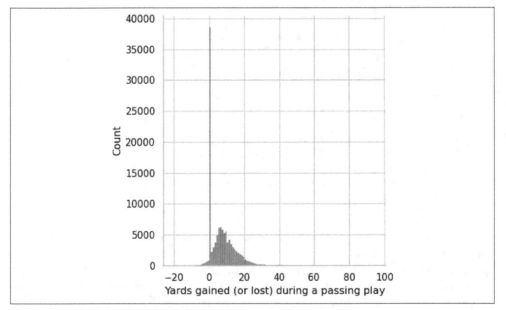

Figure 2-3. Refined histogram in Python using seaborn for the passing_yards variable

In R, filter out the long passes and make the plot look better by adding labels to the x- and y-axes and using the black-and-white theme (theme_bw()), creating Figure 2-4:

```R
## R
pbp_r_p |>
    filter(pass_length_air_yards == "long") |>
    ggplot(aes(passing_yards)) +
    geom_histogram(binwidth = 1) +
    ylab("Count") +
    xlab("Yards gained (or lost) during passing plays on long passes") +
    theme_bw()
```

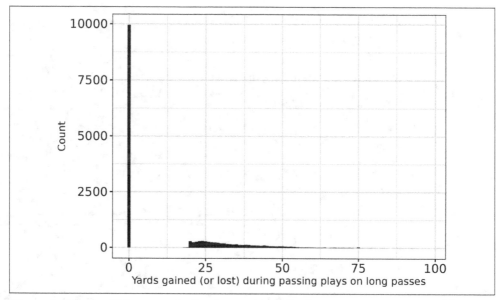

Figure 2-4. Refined histogram in R using ggplot2 for the passing_yards variable

> We will use the black-and-white theme, theme_bw(), for the remainder of the book as the default for R plots, and sns.set_theme(style="whitegrid", palette="colorblind") for Python plots. We like these themes because we think they look better on paper.

These histograms represent pictorially what you saw numerically in "Summarizing Data" on page 25. Specifically, shorter passes have fewer variable outcomes than longer passes. You can do the same thing with EPA and find similar results. For the rest of the chapter, we will stick with passing YPA for our examples, with EPA per pass attempt included within the exercises for you to do on your own.

Notice that `ggplot2`, and, more broadly, R, work well with piping objects and avoids intermediate objects. In contrast, Python works well by saving intermediate objects. Both approaches have trade-offs. For example, saving intermediate objects allows you to see the output of intermediate steps of your plotting. In contrast, rewriting the same object name can be tedious. These contrasting approaches represent a philosophical difference between the two languages. Neither is inherently right or wrong, and both have trade-offs.

Boxplots

Histograms allow people to *see* the distribution of data points. However, histograms can be cumbersome, especially when exploring many variables. Boxplots are a compromise between histograms and numerical summaries (see Table 2-1 for the numerical values). *Boxplots* get their name from the rectangular *box* containing the middle 50% of the sorted data; the line in the middle of the box is the median, so half of the sorted data falls above the line, and half of the data falls under the line.

Some people call boxplots *box-and-whisker plots* because lines extend above and under the box. These whiskers contain the remainder of the data other than outliers. Boxplots in both `seaborn` and `ggplot` use a default for *outliers* to be points that are more than 1.5 times the *interquartile range* (the range between the 25th and 75th percentiles)—either greater than the third quartile or less than the first quartile. These outliers are plotted with dots.

Table 2-1. Parts of a boxplot

Part name	Range of data
Top dots	Outliers above the data
Top whisker	100% to 75% of data, excluding outliers
Top portion of the box	75% to 50% of data
Line in the middle of the box	50% of data
Bottom portion of the box	50% to 25% of data
Bottom whisker	25% to 0% of data, excluding outliers
Bottom dots	Outliers under the data

Various types of outliers exist. Outliers may be problem data points (for example, somebody entered –10 yards when they meant 10 yards), but they often exist as parts of the data. Understanding the reasons behind these data points often provides keen insights to the data because outliers reflect the best and worst outcomes and may have interesting stories behind the points. Unless outliers exist because of errors (such as the wrong data being entered), outliers usually should be included in the data used to train models.

We place a semicolon (;) after the Python plot commands to suppress text descriptions of the plot. These semicolons are optional and simply a preference of the authors.

In Python, use the `boxplot()` function from **seaborn** and change the axes labels to create Figure 2-5:

```python
## Python
pass_boxplot = \
    sns.boxplot(data=pbp_py_p,
                x="pass_length_air_yards",
                y="passing_yards");
pass_boxplot.set(
    xlabel="Pass length (long >= 20 yards, short < 20 yards)",
    ylabel="Yards gained (or lost) during a passing play",
);
plt.show();
```

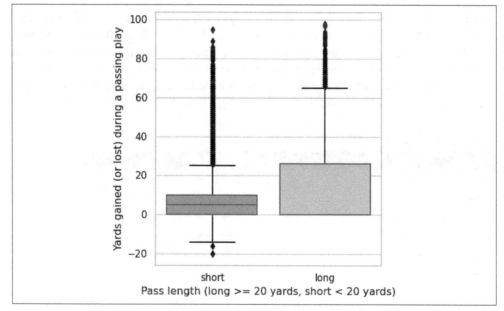

Figure 2-5. Boxplot of yards gained from long and short air-yard passes (seaborn)

In R, use `geom_boxplot()` with **ggplot2** to create Figure 2-6:

```r
## R
ggplot(pbp_r_p, aes(x = pass_length_air_yards, y = passing_yards)) +
    geom_boxplot() +
    theme_bw() +
    xlab("Pass length in yards (long >= 20 yards, short < 20 yards)") +
    ylab("Yards gained (or lost) during a passing play")
```

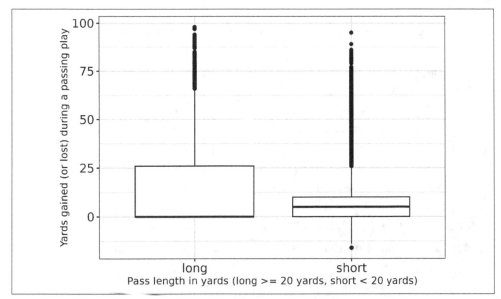

Figure 2-6. Boxplot of yards gained from long and short air-yard passes (`ggplot2`)

Player-Level Stability of Passing Yards per Attempt

Now that you've become acquainted with our data, it's time to use it for player evaluation. The first thing you have to do is aggregate across a prespecified time frame to get a value for each player. While week-level outputs certainly matter, especially for fantasy football and betting (see Chapter 7), most of the time when teams are thinking about trying to acquire a player, they use season-level data (or sometimes data over many seasons).

Thus, you aggregate at the season level here, by using the grouby() syntax in Python or group_by() syntax in R. The *group by* concept borrows from SQL-type database languages. When thinking about the process here, *group by* may be thought of as a verb. For example, you use the play-by-play data and then *group by* the *seasons* and then *aggregate* (in Python) or *summarize* (in R) to calculate the mean of the quarterback's passing YPA.

For this problem, take the play-by-play dataframe (pbp_py or pbp_r) and then *group by* passer_player_name, passer_player_id, season, and pass_length. Group by both the player ID and the player name column because some players have the same name (or at least same first initial and last name), but the name is important for studying the results of the analysis. Start with the whole dataset first before transitioning to the subsets.

In Python, use `groupby()` with a list of the variables (`["item1", "item2"]` in Python syntax) that you want to group by. Then aggregate the data for `passing_yards` for the mean:

```python
## Python
pbp_py_p_s = \
    pbp_py_p\
    .groupby(["passer_id", "passer", "season"])\
    .agg({"passing_yards": ["mean", "count"]})
```

With Python, also collapse the columns to make the dataframe easier to handle (`list()` creates a list, `map()` iterates over items, like a `for` loop without the loop syntax—see Chapter 7 details on `for` loops):

```python
## Python
pbp_py_p_s.columns = list(map("_".join, pbp_py_p_s.columns.values))
```

Next, rename the columns to names that are shorter and more intuitive:

```python
pbp_py_p_s \
    .rename(columns={'passing_yards_mean': 'ypa',
                     'passing_yards_count': 'n'},
            inplace=True)
```

In R, pipe `pbp_p` to the `group_by()` function and then use the `summarize()` function to calculate the `mean()` of `passing_yards`, as well as to calculate the number, `n()` of passing attempts for each player in each season. Include `.groups = "drop"` to tell R to drop the groupings from the resulting dataframe. The resulting mean of `passing_yards` is the YPA, which is a quarterback's average passing distance per play. Use the `<-` function to save the resulting calculations as a new dataframe, `pbp_r_p_s`:

```r
## R
pbp_r_p_s <-
    pbp_r_p |>
    group_by(passer_player_name, passer_player_id, season) |>
    summarize(
        ypa = mean(passing_yards, na.rm = TRUE),
        n = n(),
        .groups = "drop"
    )
```

Now look at the top of the resulting dataframe by using `head()` in Python and then `sort()` by `ypa` to help you better see the results. The `ascending=False` option tells Python to sort high to low (for example, arranging the values as 9, 8, 7) rather than low to high (for example, arranging the values as 7, 8, 9):

```python
## Python
pbp_py_p_s\
    .sort_values(by=["ypa"], ascending=False)\
    .head()
```

Resulting in:

```
                              ypa  n
passer_id  passer     season
00-0035544 T.Kennedy  2021    75.0  1
00-0033132 K.Byard    2018    66.0  1
00-0031235 O.Beckham  2018    53.0  2
00-0030669 A.Wilson   2018    52.0  1
00-0029632 M.Sanu     2017    51.0  1
```

In R, use `arrange()` with ypa to sort the outputs. The negative sign (–) tells R to reverse the order (for example, 7, 8, 9 becomes 9, 8, 7 when sorted):

```
## R
pbp_r_p_s |>
    arrange(-ypa) |>
    print()
```

Resulting in:

```
# A tibble: 746 × 5
   passer_player_name passer_player_id season   ypa     n
   <chr>              <chr>             <dbl> <dbl> <int>
 1 T.Kennedy          00-0035544         2021    75     1
 2 K.Byard            00-0033132         2018    66     1
 3 O.Beckham          00-0031235         2018    53     2
 4 A.Wilson           00-0030669         2018    52     1
 5 M.Sanu             00-0029632         2017    51     1
 6 C.McCaffrey        00-0033280         2018    50     1
 7 W.Snead            00-0030663         2016    50     1
 8 T.Boyd             00-0033009         2021    46     1
 9 R.Golden           00-0028954         2017    44     1
10 J.Crowder          00-0031941         2020    43     1
# i 736 more rows
```

Now this isn't really informative yet, since the players with the highest YPA values are players who threw a pass or two (usually a trick play) that were completed for big yardage. Fix this by filtering for a certain number of passing attempts in a season (let's say 100) and see what you get.

Appendix C contains more tips and tricks for data wrangling if you need more help understanding what is going on with this code. In Python, reuse pbp_py_p_s and the previous code, but include a `query()` for players with 100 or more pass attempts by using `'n >= 100'`:

```
## Python
pbp_py_p_s_100 = \
    pbp_py_p_s\
    .query("n >= 100")\
    .sort_values(by=["ypa"], ascending=False)
```

Now, look at the head of the data:

```Python
## Python
pbp_py_p_s_100.head()
```

Resulting in:

```
                                  ypa    n
passer_id passer        season
00-0023682 R.Fitzpatrick 2018    9.617886  246
00-0026143 M.Ryan        2016    9.442155  631
00-0029701 R.Tannehill   2019    9.069971  343
00-0033537 D.Watson      2020    8.898524  542
00-0036212 T.Tagovailoa  2022    8.892231  399
```

In R, *group by* the same variables and then *summarize*—this time including the number of observations per group with n(). Keep piping the results and *filter* for passers with 100 or more (n >= 100) passes and *arrange* the output:

```R
## R
pbp_r_p_100 <-
    pbp_r_p |>
    group_by(passer_id, passer, season) |>
    summarize(
        n = n(), ypa = mean(passing_yards),
        .groups = "drop"
    ) |>
    filter(n >= 100) |>
    arrange(-ypa)
```

Then print the top 20 results:

```R
## R
pbp_r_p_100 |>
    print(n = 20)
```

Which results in:

```
# A tibble: 300 × 5
   passer_id  passer        season     n   ypa
   <chr>      <chr>          <dbl> <int> <dbl>
 1 00-0023682 R.Fitzpatrick   2018   246  9.62
 2 00-0026143 M.Ryan          2016   631  9.44
 3 00-0029701 R.Tannehill     2019   343  9.07
 4 00-0033537 D.Watson        2020   542  8.90
 5 00-0036212 T.Tagovailoa    2022   399  8.89
 6 00-0031345 J.Garoppolo     2017   176  8.86
 7 00-0033873 P.Mahomes       2018   651  8.71
 8 00-0036442 J.Burrow        2021   659  8.67
 9 00-0026498 M.Stafford      2019   289  8.65
10 00-0031345 J.Garoppolo     2021   511  8.50
11 00-0033319 N.Mullens       2018   270  8.43
12 00-0033537 D.Watson        2017   202  8.41
13 00-0033077 D.Prescott      2020   221  8.40
```

```
14 00-0034869 S.Darnold      2022  137  8.34
15 00-0037834 B.Purdy        2022  233  8.34
16 00-0029604 K.Cousins      2020  513  8.31
17 00-0031345 J.Garoppolo    2019  532  8.28
18 00-0025708 M.Moore        2016  122  8.28
19 00-0033873 P.Mahomes      2019  596  8.28
20 00-0020531 D.Brees        2017  606  8.26
# i 280 more rows
```

Even the most astute of you probably didn't expect the Harvard-educated Ryan Fitzpatrick's season as Jameis Winston's backup to appear at the top of this list. You do see the MVP seasons of Matt Ryan (2016) and Patrick Mahomes (2018), and a bunch of quarterbacks (including Matt Ryan) coached by the great Kyle Shanahan.

Deep Passes Versus Short Passes

Now, down to the business of the chapter, testing the second part of the hypothesis: "Throwing deep passes is more valuable than short passes, but it's difficult to say whether or not a quarterback is good at deep passes." For this stability analysis, do the following steps:

1. Calculate the YPA for each passer for each season.

2. Calculate the YPA for each passer for the previous season.

3. Look at the correlation from the values calculated in steps 1 and 2 to see the stability.

Use similar code as before, but include `pass_length_air_yards` with the *group by* commands to include pass yards. With this operation, naming becomes hard.

We have you use the dataset (*play-by-play*, pbp), the language (either Python, _py, or R, _r), passing plays (_p), seasons data (_s), and finally, pass length (_pl).

For both languages, you will create a copy of the dataframe and then shift the year by adding 1. Then you'll merge the new dataframe with the original dataframe. This will let you have the current and previous year's values.

 Longer names are tedious, but we have found unique names to be important so that you can quickly search through code by using tools like Find and Replace (which are found in most code editors) to see what is occurring with your code (with Find) or change names (with Replace).

In Python, create pbp_r_p_s_pl, using several steps. First, *group by* and *aggregate* to get the mean and count:

```Python
## Python
pbp_py_p_s_pl = \
    pbp_py_p\
    .groupby(["passer_id", "passer", "season", "pass_length_air_yards"])\
    .agg({"passing_yards": ["mean", "count"]})
```

Next, flatten the column names and rename passing_yards_mean to **ypa** and pass ing_yards_count to **n** in order to have shorter names that are easier to work with:

```Python
## Python
pbp_py_p_s_pl.columns =\
    list(map("_".join, pbp_py_p_s_pl.columns.values))
pbp_py_p_s_pl\
    .rename(columns={'passing_yards_mean': 'ypa',
                     'passing_yards_count': 'n'},
            inplace=True)
```

Next, reset the index:

```Python
## Python
pbp_py_p_s_pl.reset_index(inplace=True)
```

Select only short-passing data from passers with more than 100 such plays and long-passing data for passers with more than 30 such plays:

```Python
## Python
q_value = (
    '(n >= 100 & ' +
    'pass_length_air_yards == "short") | ' +
    '(n >= 30 & ' +
    'pass_length_air_yards == "long")'
)
pbp_py_p_s_pl = pbp_py_p_s_pl.query(q_value).reset_index()
```

Then create a list of columns to save (cols_save) and a new dataframe with only these columns (air_yards_py). Include a .copy() so edits will not be passed back to the original dataframe:

```Python
## Python
cols_save =\
    ["passer_id", "passer", "season",
     "pass_length_air_yards", "ypa"]
air_yards_py =\
    pbp_py_p_s_pl[cols_save].copy()
```

Next, copy `air_yards_py` to create `air_yards_lag_py`. Take the current season value and add 1 by using the shortcut command += and rename the `passing_yards_mean` to include `lag` (which refers to the one-year offset or delay between the two years):

```Python
## Python
air_yards_lag_py =\
    air_yards_py\
    .copy()
air_yards_lag_py["season"] += 1
air_yards_lag_py\
    .rename(columns={'ypa': 'ypa_last'},
    inplace=True)
```

Finally, `merge()` the two dataframes together to create `air_yards_both_py` and use an *inner join* so only shared years will be saved and join *on* `passer_id`, `passer`, season, and `pass_length_air_yards`:

```Python
## Python
pbp_py_p_s_pl =\
    air_yards_py\
    .merge(air_yards_lag_py,
        how='inner',
        on=['passer_id', 'passer',
            'season', 'pass_length_air_yards'])
```

Check the results of your choice in Python by examining a couple of quarterbacks of your choice such as Tom Brady (`T.Brady`) and Aaron Rodgers (`A.Rodgers`) and include only the necessary columns to have an easier-to-view dataframe:

```Python
## Python
print(
    pbp_py_p_s_pl[["pass_length_air_yards", "passer",
        "season", "ypa", "ypa_last"]]\
    .query('passer == "T.Brady" | passer == "A.Rodgers"')\
    .sort_values(["passer", "pass_length_air_yards", "season"])\
    .to_string()
)
```

Resulting in:

```
    pass_length_air_yards    passer  season        ypa   ypa_last
47                   long  A.Rodgers   2019  12.092593  12.011628
49                   long  A.Rodgers   2020  16.097826  12.092593
51                   long  A.Rodgers   2021  14.302632  16.097826
53                   long  A.Rodgers   2022  10.312500  14.302632
45                  short  A.Rodgers   2017   6.041475   6.693523
46                  short  A.Rodgers   2018   6.697446   6.041475
48                  short  A.Rodgers   2019   6.207224   6.697446
50                  short  A.Rodgers   2020   6.718447   6.207224
52                  short  A.Rodgers   2021   6.777083   6.718447
54                  short  A.Rodgers   2022   6.239130   6.777083
0                    long    T.Brady   2017  13.264706  15.768116
```

2	long	T.Brady	2018	10.232877	13.264706
4	long	T.Brady	2019	10.828571	10.232877
6	long	T.Brady	2020	12.252101	10.828571
8	long	T.Brady	2021	12.242424	12.252101
10	long	T.Brady	2022	10.802469	12.242424
1	short	T.Brady	2017	7.071429	7.163022
3	short	T.Brady	2018	7.356452	7.071429
5	short	T.Brady	2019	6.048276	7.356452
7	short	T.Brady	2020	6.777600	6.048276
9	short	T.Brady	2021	6.634697	6.777600
11	short	T.Brady	2022	5.832168	6.634697

 We suggest using at least two players to check your code. For example, Tom Brady is the first player by passer_id, and looking at only his values might not show a mistake that does not affect the first player in the dataframe.

In R, similar steps are taken to create `pbp_r_p_s_pl`. First, create `air_yards_r` by selecting the columns needed and arrange the dataframe:

```
## R
air_yards_r <-
    pbp_r_p |>
    select(passer_id, passer, season,
           pass_length_air_yards, passing_yards) |>
    arrange(passer_id, season,
            pass_length_air_yards) |>
    group_by(passer_id, passer,
             pass_length_air_yards, season) |>
    summarize(n = n(),
              ypa = mean(passing_yards),
              .groups = "drop") |>
    filter((n >= 100 & pass_length_air_yards == "short") |
           (n >= 30 & pass_length_air_yards == "long")) |>
    select(-n)
```

Next, create the lag dataframe including a mutate to the seasons and add 1:

```
## R
air_yards_lag_r <-
    air_yards_r |>
    mutate(season = season + 1) |>
    rename(ypa_last = ypa)
```

Last, join the dataframes to create `pbp_r_p_s_pl`:

```
## R
pbp_r_p_s_pl <-
    air_yards_r |>
    inner_join(air_yards_lag_r,
```

```
        by = c("passer_id", "pass_length_air_yards",
                "season", "passer"))
```

Check the results in R by examining passers of your choice such as Tom Brady (T.Brady) and Aaron Rodgers (A.Rodgers):

```
## R
pbp_r_p_s_pl |>
    filter(passer %in% c("T.Brady", "A.Rodgers")) |>
    print(n = Inf)
```

Which results in:

```
# A tibble: 22 × 6
   passer_id  passer    pass_length_air_yards season   ypa ypa_last
   <chr>      <chr>     <chr>                  <dbl> <dbl>    <dbl>
 1 00-0019596 T.Brady   long                    2017  13.3     15.8
 2 00-0019596 T.Brady   long                    2018  10.2     13.3
 3 00-0019596 T.Brady   long                    2019  10.8     10.2
 4 00-0019596 T.Brady   long                    2020  12.3     10.8
 5 00-0019596 T.Brady   long                    2021  12.2     12.3
 6 00-0019596 T.Brady   long                    2022  10.8     12.2
 7 00-0019596 T.Brady   short                   2017  7.07     7.16
 8 00-0019596 T.Brady   short                   2018  7.36     7.07
 9 00-0019596 T.Brady   short                   2019  6.05     7.36
10 00-0019596 T.Brady   short                   2020  6.78     6.05
11 00-0019596 T.Brady   short                   2021  6.63     6.78
12 00-0019596 T.Brady   short                   2022  5.83     6.63
13 00-0023459 A.Rodgers long                    2019  12.1     12.0
14 00-0023459 A.Rodgers long                    2020  16.1     12.1
15 00-0023459 A.Rodgers long                    2021  14.3     16.1
16 00-0023459 A.Rodgers long                    2022  10.3     14.3
17 00-0023459 A.Rodgers short                   2017  6.04     6.69
18 00-0023459 A.Rodgers short                   2018  6.70     6.04
19 00-0023459 A.Rodgers short                   2019  6.21     6.70
20 00-0023459 A.Rodgers short                   2020  6.72     6.21
21 00-0023459 A.Rodgers short                   2021  6.78     6.72
22 00-0023459 A.Rodgers short                   2022  6.24     6.78
```

We use the philosophy "Assume your code is wrong until you have convinced yourself it is correct." Hence, we often peek at our code to make sure we understand what the code is doing versus what we think the code is doing. Practically, this means following the advice of former US President Ronald Reagan: "Trust but verify" your code.

The dataframes you've created (either pbp_py_p_s_pl in Python or pbp_r_p_s_pl in R) now contain six columns. Look at the info() for the dataframe in Python:

```
## Python
pbp_py_p_s_pl\
    .info()
```

Resulting in:

```
<class 'pandas.core.frame.DataFrame'>
RangeIndex: 317 entries, 0 to 316
Data columns (total 6 columns):
 #   Column                Non-Null Count  Dtype
---  ------                --------------  -----
 0   passer_id             317 non-null    object
 1   passer                317 non-null    object
 2   season                317 non-null    int64
 3   pass_length_air_yards 317 non-null    object
 4   ypa                   317 non-null    float64
 5   ypa_last              317 non-null    float64
dtypes: float64(2), int64(1), object(3)
memory usage: 15.0+ KB
```

Or glimpse() at the dataframe in R:

```
## R
pbp_r_p_s_pl |>
    glimpse()
```

Resulting in:

```
Rows: 317
Columns: 6
$ passer_id            <chr> "00-0019596", "00-0019596", "00-0019596", "00-00…
$ passer               <chr> "T.Brady", "T.Brady", "T.Brady", "T.Brady", "T.B…
$ pass_length_air_yards <chr> "long", "long", "long", "long", "long", "long", …
$ season               <dbl> 2017, 2018, 2019, 2020, 2021, 2022, 2017, 2018, …
$ ypa                  <dbl> 13.264706, 10.232877, 10.828571, 12.252101, 12.2…
$ ypa_last             <dbl> 15.768116, 13.264706, 10.232877, 10.828571, 12.2…
```

The six columns contain the following data:

- passer_id is the unique passer identification number for the player.

- passer is the (potentially) nonunique first initial and last name for the passer.

- pass_length_air_yards is the type of pass (either long or short) you defined earlier.

- season is the final season in the season pair (e.g., season being 2017 means you're comparing 2016 and 2017).

- ypa is the yards per attempt during the stated season (e.g., 2017 in the previous example).

- ypa_last is the yards per attempt during the season previous to the stated season (e.g., 2016 in the previous example).

Now that we've reminded ourselves what's in the data, let's dig in and see how many quarterbacks you have. With Python, use the `passer_id` column and find the `unique()` values and then find the length of this object:

```
## Python
len(pbp_py_p_s_pl.passer_id.unique())
```

Resulting in:

```
65
```

With R, use the (distinct) function with passer_id and then see how many rows exist:

```
## R
pbp_r_p_s_pl |>
    distinct(passer_id) |>
    nrow()
```

Resulting in:

```
[1] 65
```

You now have a decent sample size of quarterbacks. You can plot this data by using a scatterplot. *Scatterplots* plot points on a figure, which is in contrast to histograms that plot bins of data and to boxplots that plot summaries of the data such as medians. Scatterplots allow you to "see" the data directly. The horizontal axis is called the *x-axis* and typically includes the predictor, or causal, variable, if one exists. The vertical axis is called the *y-axis* and typically includes the response, or effect, variables, if one exists. With our example, you will use the YPA from the previous year as the predictor for YPA in the current year. Plot this in R by using `geom_point()` and call this plot `scatter_ypa_r` and then print `scatter_ypa_r` to create Figure 2-7:

```
## R
scatter_ypa_r <-
    ggplot(pbp_r_p_s_pl, aes(x = ypa_last, y = ypa)) +
    geom_point() +
    facet_grid(cols = vars(pass_length_air_yards)) +
    labs(
        x = "Yards per Attempt, Year n",
        y = "Yards per Attempt, Year n + 1"
    ) +
    theme_bw() +
    theme(strip.background = element_blank())

print(scatter_ypa_r)
```

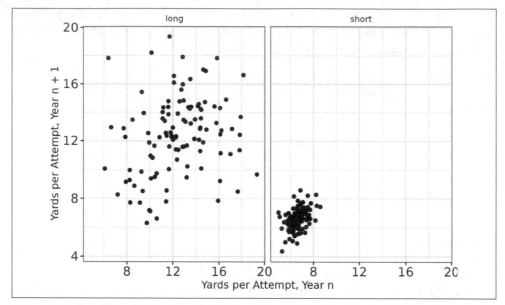

Figure 2-7. Stability of YPA plotted with ggplot2. *Notice that both sub-plots have the same x and y scales*

Figure 2-7 is encouraging for short passes. It appears that quarterbacks who are good on short passes in one year are good the following year, and vice versa. Notice that the long passes are much more unwieldy. To help you better examine these trends, include a line of best fit to the data (this is why we had you save scatter_ypa_r, so that you could reuse the plot here) to create Figure 2-8:

```
## R
# add geom_smooth() to the previously saved plot
scatter_ypa_r +
    geom_smooth(method = "lm")
```

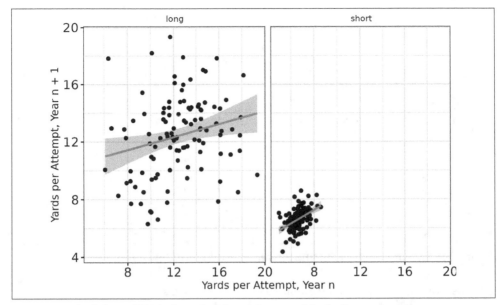

Figure 2-8. Stability of YPA plotted with `ggplot2` *and including a trend line*

For both pass types, the lines in Figure 2-8 have a slightly positive slope (the lines are increasing across the plot), but this is hard to see. To obtain this estimate using the correlations, look at the numerical values:

```
## R
pbp_r_p_s_pl |>
    filter(!is.na(ypa) & !is.na(ypa_last)) |>
    group_by(pass_length_air_yards) |>
    summarize(correlation = cor(ypa, ypa_last))
```

Resulting in:

```
# A tibble: 2 × 2
  pass_length_air_yards correlation
  <chr>                       <dbl>
1 long                        0.234
2 short                       0.438
```

These figures and analyses may be repeated in Python to create Figure 2-9:

```
## Python
sns.lmplot(data=pbp_py_p_s_pl,
           x="ypa",
           y="ypa_last",
           col="pass_length_air_yards");
plt.show();
```

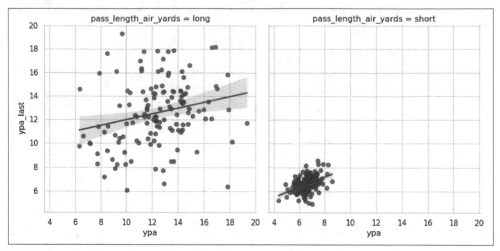

*Figure 2-9. Stability of YPA plotted with *seaborn* and including a trend line*

Likewise, the correlation can be obtained by using `pandas` as well:

```Python
## Python
pbp_py_p_s_pl\
    .query("ypa.notnull() & ypa_last.notnull()")\
    .groupby("pass_length_air_yards")[["ypa", "ypa_last"]]\
    .corr()
```

Resulting in:

```
                                ypa  ypa_last
pass_length_air_yards
long                  ypa      1.000000  0.233890
                      ypa_last 0.233890  1.000000
short                 ypa      1.000000  0.438479
                      ypa_last 0.438479  1.000000
```

The Pearson's correlation coefficient numerically captures what Figures 2-8 and 2-9 show.

While both datasets include a decent amount of noise, vis-à-vis Pearson's correlation coefficient, a quarterback's performance on shorter passes is twice as stable as on longer passes. Thus, you can confirm the second part of the guiding hypothesis of the chapter: "Throwing deep passes is more valuable than short passes, but it's difficult to say whether or not a quarterback is good at deep passes."

A Pearson's correlation coefficient can vary from –1 to 1. In the case of stability, a number closer to +1 implies strong, positive correlations and more stability, and a number closer to 0 implies weak correlations at best (and an unstable measure). A Pearson's correlation coefficient of –1 implies a decreasing correlation and does not exist for stability but would mean a high value this year would be correlated with a low value next year.

So, What Should We Do with This Insight?

Generally speaking, noisy data is a place to look for players (or teams or units within teams) that have pop-up seasons that are not likely to repeat themselves. A baseball player who sees a 20-point jump in his average based on a higher *batting average on balls in play* (*BABIP*) one year might be someone you want to avoid rostering in fantasy or real baseball. Similarly, a weaker quarterback who generates a high YPA (or EPA per pass attempt) on deep passes one year—without a corresponding increase in such metrics on shorter passes, the more stable of the two—might be what analysts call a *regression candidate*.

For example, let's look at the leaderboard for 2017 deep passing YPA in Python:

```python
## Python
pbp_py_p_s_pl\
    .query(
        'pass_length_air_yards == "long" & season == 2017'
        )[["passer_id", "passer", "ypa"]]\
    .sort_values(["ypa"], ascending=False)\
    .head(10)
```

Resulting in:

```
     passer_id      passer        ypa
41   00-0023436     A.Smith    19.338235
79   00-0026498   M.Stafford   17.830769
12   00-0020531     D.Brees    16.632353
191  00-0032950     C.Wentz    13.555556
33   00-0022942    P.Rivers    13.347826
0    00-0019596     T.Brady    13.264706
129  00-0029604   K.Cousins    12.847458
114  00-0029263    R.Wilson    12.738636
203  00-0033077   D.Prescott   12.585366
109  00-0028986    C.Keenum    11.904762
```

Some good names are on this list (Drew Brees, Tom Brady, Russell Wilson) but also some so-so names. Let's look at the same list in 2018:

```Python
## Python
pbp_py_p_s_pl\
    .query(
        'pass_length_air_yards == "long" & season == 2018'
        )[["passer_id", "passer", "ypa"]]\
    .sort_values(["ypa"], ascending=False)\
    .head(10)
```

Resulting in:

```
        passer_id       passer       ypa
116     00-0029263    R.Wilson  15.597403
14      00-0020531     D.Brees  14.903226
205     00-0033077  D.Prescott  14.771930
214     00-0033106      J.Goff  14.445946
35      00-0022942    P.Rivers  14.357143
157     00-0031280      D.Carr  14.339286
188     00-0032268   M.Mariota  13.941176
64      00-0026143      M.Ryan  13.465753
193     00-0032950     C.Wentz  13.222222
24      00-0022803   E.Manning  12.941176
```

Alex Smith, who was long thought of as a dink-and-dunk specialist, dropped off this list completely. He led the league in passer rating in 2017, before being traded by Kansas City to Washington for a third-round pick and star cornerback Kendall Fuller (there's a team that knows how to sell high!).

While the list includes some repeats or YPA on deep passes, many new names emerge. Specifically, if you filter for Matt Ryan's name in the dataset, you'll find that he averaged 17.7 YPA on deep passes in 2016 (when he won NFL MVP). In 2017, that value fell to 8.5, then rose back up to 13.5 in 2018. Did Ryan's ability drastically change during these three years, or was he subject to significant statistical variability? The math would suggest the latter. In fantasy football or betting, he would have been a *sell-high* candidate in 2017 and a *buy-low* candidate in 2018 as a result.

Data Science Tools Used in This Chapter

This chapter covered the following topics:

- Obtaining data from multiple seasons by using the nflfastR package either directly in R or via the nfl_data_py package in Python
- Changing columns based on conditions by using where in Python or ifelse() statements in R
- Using describe() for data with pandas or summarize() for data in R

- Reordering values by using `sort_by()` in Python or `arrange()` in R
- Calculating the difference between years by using `merge()` in Python or `join()` in R

Exercises

1. Create the same histograms as in "Histograms" on page 30 but for EPA per pass attempt.

2. Create the same boxplots as in "Histograms" on page 30 but for EPA per pass attempt.

3. Perform the same stability analysis as in "Player-Level Stability of Passing Yards per Attempt" on page 37, but for EPA per pass attempt. Do you see the same qualitative results as when you use YPA? Do any players have similar YPA numbers one year to the next but have drastically different EPA per pass attempt numbers across years? Where could this come from?

4. One of the reasons that data for long pass attempts is less stable than short pass attempts is that there are fewer of them, which is largely a product of 20 yards being an arbitrary cutoff for long passes (by companies like PFF). Find a cutoff that equally splits the data and perform the same analysis. Do the results stay the same?

Suggested Readings

If you want to learn more about plotting, here are some resources that we found helpful:

- *The Visual Display of Quantitative Information* by Edward Tufte (*https://oreil.ly/BYBhX*) (Graphics Press, 2001). This book is a classic on how to think about data. The book does not contain code but instead shows how to see information for data. The guidance in the book is priceless.

- The `ggplot2` package documentation (*https://ggplot2.tidyverse.org*). For those of you using R, this is the place to start to learn more about `ggplot2`. The page includes beginner resources and links to advanced resources. The page also includes examples that are great to browse.

- The `seaborn` package documentation (*https://seaborn.pydata.org*). For those of you using Python, this is the place to start for learning more about `seaborn`. The page includes beginner resources and links to advanced resources. The page also includes examples that are great to browse. The gallery on this page is especially helpful when trying to think about how to visualize data.

- *ggplot2: Elegant Graphics for Data Analysis* (*https://ggplot2-book.org*), 3rd edition, by Hadley Wickham et al. (Springer). The third edition is currently under development and accessible online. This book explains how `ggplot2` works in great detail but also provides a good method for thinking about plotting data using words. You can become an expert in `ggplot2` by reading this book while analyzing and tweaking each line of code presented. But this is not necessarily an easy route.

## CHAPTER 3

Simple Linear Regression: Rushing Yards Over Expected

Football is a contextual sport. Consider whether a pass is completed. This depends on multiple factors: Was the quarterback under pressure (making it harder to complete)? Was the defense expecting a pass (which would make it harder to complete)? What was the depth of the pass (completion percentage goes down with the depth of the target)?

What turns people off to football analytics are conclusions that they feel lack a contextual understanding of the game. "Raw numbers" can be misleading. Sam Bradford once set the NFL record for completion percentage in a season as a member of the Minnesota Vikings in 2016. This was impressive, since he joined the team early in the season as a part of a trade and had to acclimate quickly to a new environment. While that was impressive, it did not necessarily mean he was the best quarterback in the NFL that year, or even the most accurate one. For one, he averaged just 6.6 yards average depth of target (aDOT) that year, which was 37th in the NFL according to PFF. That left his yards per pass attempt at a relatively average 7.0, tied for just 20th in football. Chapter 4 provides more context for that number and shows you how to adjust it yourself.

Luckily, given the great work of the people supporting `nflfastR`, you can provide your own context for metrics by applying the statistical tool known as *regression*. Through regression, you can *normalize*, or *control for*, variables (or *features*) that have been shown to affect a player's production. Whether a feature predicts a player's production is incredibly hard to prove in real life. Also, players emerge who come along and challenge our assumptions in this regard (such as Patrick Mahomes of the Kansas City Chiefs or Derrick Henry of the Tennessee Titans). Furthermore, data often fails to capture many factors that affect performance. As in life, you cannot account for everything, but hopefully you can capture the most important things. One of Richard's professors at Texas Tech University, Katharine Long, likes to define this approach as the Mick Jagger theorem: "You can't always get what you want, but if you try sometimes, you just might get what you need."

The process of normalization in both the public and private football analytics space generally requires models that are more involved than a simple linear regression, the model covered in this chapter. But we have to start somewhere. And simple linear regression provides a nice start to modeling because it is both understandable and the foundation for many other types of analyses.

 Many fields use simple linear regression, which leads to the use of multiple terms. Mathematically, the predictor variable is usually x, and the response variable is usually y. Some synonyms for x include *predictor variable*, *feature*, *explanatory variable*, and *independent variable*. Some synonyms for y include *response variable*, *target*, and *dependent variable*. Likewise, medical studies often *correct for* exogenous or confounding data (*variables* to statisticians or *features* to data scientists) such as education level, age, or other socioeconomic data. You are learning the same concepts in this chapter and Chapter 4 with the terms *normalize* and *control for*.

Simple linear regression consists of a model with a single explanatory variable that is assumed to be linearly related to a single dependent variable, or *feature*. A *simple linear regression* fits the statistically "best" straight line by using one independent predictor variable to estimate a response variable as a function of the predictor. *Simple* refers to having only one predictor variable as well an intercept, an assumption Chapter 4 shows you how to relax. *Linear* refers to the straight line (compared to a curved line or polynomial line for those of you who remember high school algebra).

Regression originally referred to the idea that observations will return, or *regress*, to the average over time, as noted by Francis Galton in 1877 (*https://oreil.ly/5hyWI*). For example, if a running back has above-average rushing yards per carry one year, we would statistically expect them to revert, or *regress*, to the league average in future years, all else being equal. The linear assumption made in many models is often onerous but is generally fine as a first pass.

To start applying simple linear regression, you are going to work on a problem that has been solved already in the public space, during the 2020 Big Data Bowl. Participants in this event used *tracking data* (the positioning, direction, and orientation of all 22 on-field players every tenth of a second) to model the expected rushing yards gained on a play. This value was then subtracted from a player's actual rushing yards on a play to determine their *rushing yards over expected* (*RYOE*). As we talked about in Chapter 1, this kind of residual analysis is a cornerstone exercise in all of sports analytics.

The RYOE metric has since made its way onto broadcasts of NFL games. Additional work has been done to improve the metric, including creating a version that uses scouting data instead of tracking data, as was done by Tej Seth at PFF, using an R Shiny app for RYOE (*https://oreil.ly/ZD2V_*). Regardless of the mechanics of the model, the broad idea is to adjust for the situation a rusher has to undergo to gain yards.

 The Big Data Bowl (*https://oreil.ly/XAXTJ*) is the brainchild of Michael Lopez, the NFL's director of data and analytics. Like Eric, Lopez was previously a professor, in Lopez's case, at Skidmore College as a professor of statistics. Lopez's home page (*https://stats bylopez.com*) contains useful tips, insight, and advice for sports as well as careers.

To mimic RYOE, but on a much smaller scale, you will use the *yards to go* on a given play. Recall that each football play has a down and distance, where *down* refers to the place in the four-down sequence a team is in to either pick up 10 yards or score either a touchdown or field goal. *Distance*, or *yards to go*, refers to the distance left to achieve that goal and is coded in the data as ydstogo.

A reasonable person would expect that the specific down and yards to go affect RYOE. This observation occurs because it is easier to run the ball when more yards to go exist because the defense will usually try to prevent longer plays. For example, when an offense faces third down and 10 yards to go, the defense is playing back in hopes of avoiding a big play. Conversely, while on second down and 1 yard to go, the defense is playing up to try to prevent a first down or touchdown.

For many years, teams deployed a *short-yardage back*, a (usually larger) running back who would be tasked with gaining the (small) number of yards on third or fourth down when only 1 or 2 yards were required for a first down or touchdown. These players were prized in fantasy football for their abilities to *vulture* (or take credit for) touchdowns that a team's starting running back often did much of the work for, for their team to score. But the short-yardage backs' yards-per-carry values were not impressive compared to the starting running back. This short-yardage back's yards per carry lacked context compared to the starting running back's. Hence, metrics like RYOE help normalize the context of a running back's plays.

Many example players exist from the history of the NFL. Mike Alstott, the Tampa Bay Buccaneers second-round pick in 1996, often served as the short-yardage back on the upstart Bucs teams of the late-1990s/early-2000s. In contrast, his backfield mate, Warrick Dunn, the team's first-round choice in 1997, served in the "early-down" role.

As a result, their yards-per-carry numbers were different as members of the same team: 3.7 yards for Alstott and 4.0 yards for Dunn. Thus, regression can help you account for that and make better comparisons to create metrics such as RYOE.

The wisdom of drafting a running back in the top two rounds once, let alone in consecutive years, is a whole other topic in football analytics. We talk about the draft in great detail in Chapter 7.

Understanding simple linear regression from this chapter also serves as a foundation for skills covered in other chapters, such as more complex RYOE models in Chapter 4, completion percentage over expected in the passing game in Chapter 5, touchdown passes per game in Chapter 6, and models used to evaluate draft data in Chapter 7. Many people, including the authors, call linear models both the workhorse and foundation for applied statistics and data science.

Exploratory Data Analysis

Prior to running a simple linear regression, it's always good to plot the data as a part of the modeling process, using the exploratory data analysis (EDA) skills you learned about in Chapter 2. You will do this using seaborn in Python or ggplot2 in R. Before you calculate the RYOE, you need to load and wrangle the data. You will use the data from 2016 to 2022. First, load the packages and data.

Make sure you have installed the statsmodels package using **pip install statsmodels** in the terminal.

If you're using Python, use this code to load the data:

```Python
## Python
import pandas as pd
import numpy as np
import nfl_data_py as nfl
import statsmodels.formula.api as smf
import matplotlib.pyplot as plt
import seaborn as sns

seasons = range(2016, 2022 + 1)
pbp_py = nfl.import_pbp_data(seasons)
```

If you're using R, use this code to load the data:

```
## R
library(tidyverse)
library(nflfastR)

pbp_r <- load_pbp(2016:2022)
```

After loading the data, select the running plays. Use the filtering criteria `play_type == "run"`. Also, remove the plays without a rusher and replace the missing rushing yards with 0.

In Python, use `rusher_id.notnull()` as part of your query and then replace missing `rushing_yards` values with 0:

```
## Python
pbp_py_run =\
    pbp_py.query('play_type == "run" & rusher_id.notnull()')\
    .reset_index()
pbp_py_run\
    .loc[pbp_py_run.rushing_yards.isnull(), "rushing_yards"] = 0
```

In R, use `!is.na(rusher_id)` as part of your `filter()` step and then `mutate()` with an `ifelse()` function to replace the missing values:

```
## R
pbp_r_run <-
    pbp_r |>
    filter(play_type == "run" & !is.na(rusher_id)) |>
    mutate(rushing_yards = ifelse(is.na(rushing_yards), 0, rushing_yards))
```

Next, plot the raw data prior to building a model. In Python, use `displot()` from seaborn to create Figure 3-1:

```
## Python
sns.set_theme(style="whitegrid", palette="colorblind")
sns.scatterplot(data=pbp_py_run, x="ydstogo", y="rushing_yards");
plt.show();
```

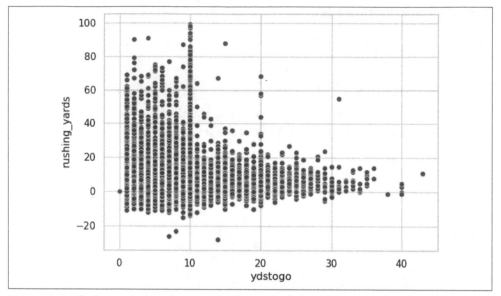

Figure 3-1. Yards to go plotted against rushing yards, using seaborn

In R, use `geom_point()` from `ggplot2` to create Figure 3-2:

```
ggplot(pbp_r_run, aes(x = ydstogo, y = rushing_yards)) +
    geom_point() +
    theme_bw()
```

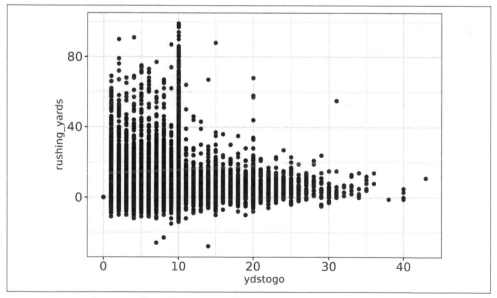

Figure 3-2. Yards to go plotted against rushing yards, using ggplot2

Figures 3-1 and 3-2 are dense graphs of points, and it's hard to see whether a relation exists between the yards to go and the number of rushing yards gained on a play. You can do some things to make the plot easier to read. First, add a trend line to see if the data slopes upward, downward, or neither up nor down.

In Python, use `regplot()` to create Figure 3-3:

```Python
## Python
sns.regplot(data=pbp_py_run, x="ydstogo", y="rushing_yards");
plt.show();
```

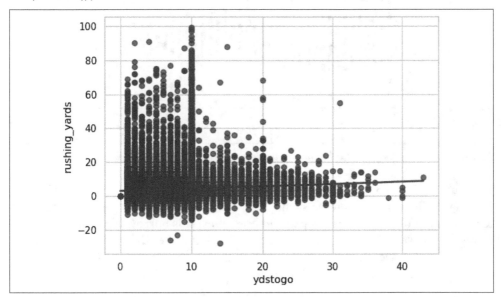

Figure 3-3. Yards to go plotted against rushing yards with a trend line (seaborn)

In R, use `stat_smooth(method = "lm")` with your code from Figure 3-2 to create Figure 3-4:

```
ggplot(pbp_r_run, aes(x = ydstogo, y = rushing_yards)) +
    geom_point() +
    theme_bw() +
    stat_smooth(method = "lm")
```

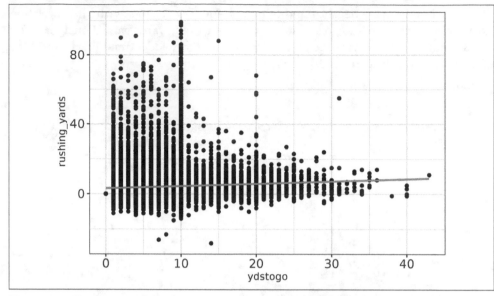

Figure 3-4. Yards to go plotted against rushing yards with a trendline (`ggplot2`)

In Figures 3-3 and 3-4, you see a positive slope, albeit a very small one. This shows you that rushing gains increase slightly as yards to go increases. Another approach to try to examine the data is *binning and averaging*. This borrows from the ideas of a histogram (covered in "Histograms" on page 30), but rather than using the count for each bin, an average is used for each bin. In this case, the bins are easy to define: they are the ydstogo values, which are integers.

Now, average over each yards-per-carry value gained in each bin. In Python, aggregate the data and then plot it to create Figure 3-5:

```python
## Python
pbp_py_run_ave =\
    pbp_py_run.groupby(["ydstogo"])\
    .agg({"rushing_yards": ["mean"]})

pbp_py_run_ave.columns = \
    list(map("_".join, pbp_py_run_ave.columns))
pbp_py_run_ave\
    .reset_index(inplace=True)

sns.regplot(data=pbp_py_run_ave, x="ydstogo", y="rushing_yards_mean");
plt.show();
```

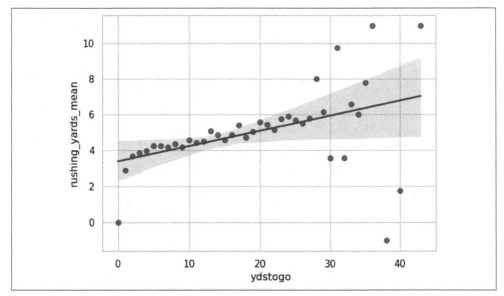

Figure 3-5. Average yards per carry plotted with seaborn

In R, create a new variable, *yards per carry* (ypc) and then plot the results in Figure 3-6:

```r
## R
pbp_r_run_ave <-
    pbp_r_run |>
    group_by(ydstogo) |>
    summarize(ypc = mean(rushing_yards))

ggplot(pbp_r_run_ave, aes(x = ydstogo, y = ypc)) +
    geom_point() +
    theme_bw() +
    stat_smooth(method = "lm")
```

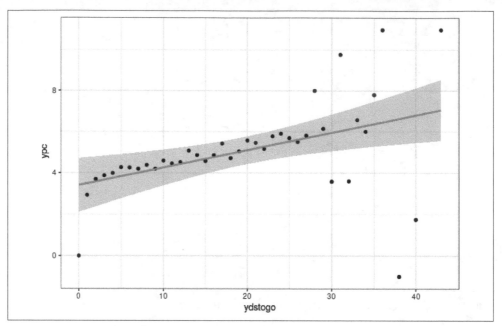

Figure 3-6. Average yards per carry plotted with `ggplot2`

 Figures 3-5 and 3-6 let you see a positive linear relationship between average yards gained and yards to go. While binning and averaging is not a substitute for regressing along the entire dataset, the approach can give you insight into whether such an endeavor is worth doing in the first place and helps you to better "see" the data.

Simple Linear Regression

Now that you've wrangled and interrogated the data, you're ready to run a simple linear regression. Python and R use the same formula notation for the functions we show you in this book. For example, to build a simple linear regression where `ydstogo` predicts `rushing_yards`, you use the formula `rushing_yards ~ 1 + ydstogo`.

 The left hand side of the formula contains the target, or response, variable. The right hand side of the formula contains the response, or predictor, variables. Chapter 4 shows how to use multiple predictors that are separated by a +.

You can read this formula as `rushing_yards` are *predicted by* (indicated by the tilde, ~, which is located next to the 1 key on US keyboards and whose location varies on other keyboards) an *intercept* (1) and a *slope parameter* for yards to go (`ydstogo`). The 1 is an optional value to explicitly tell you where the model contains an intercept. Our code and most people's code we read do not usually include an intercept in the formula, but we include it here to help you explicitly think about this term in the model.

Formulas with `statsmodels` are usually similar or identical to R. This is because computer languages often borrow from other computer languages. Python's `statsmodels` borrowed formulas from R, similar to `pandas` borrowing dataframes from R. R also borrows ideas, and R is, in fact, an open source re-creation of the S language. As another example, both the `tidyverse` in R and `pandas` in Python borrow syntax and ideas for cleaning data from SQL-type languages.

We are using the `statsmodels` package because it is better for statistical inference compared to the more popular Python package `scikit-learn` that is better for machine learning. Additionally, `statsmodels` uses similar syntax as R, which allows you to more readily compare the two languages.

In Python, use the `statsmodels` package's `formula.api` imported as `smf`, to run an *ordinary least-squares regression*, `ols()`. To build the model, you will need to tell Python how to fit the regression. For this, model that the number of rushing yards for a play (`rushing_yards`) is predicted by an intercept (1) and the number of yards to go for the play (`ydstogo`), which is written as a formula: `rushing_yards ~ 1 + ydstogo`.

Using Python, build the model, fit the model, and then look at the model's summary:

```Python
## Python
import statsmodels.formula.api as smf

yard_to_go_py =\
    smf.ols(formula='rushing_yards ~ 1 + ydstogo', data=pbp_py_run)

print(yard_to_go_py.fit().summary())
```

Resulting in:

```
                         OLS Regression Results
==============================================================================
Dep. Variable:         rushing_yards   R-squared:                       0.007
Model:                           OLS   Adj. R-squared:                  0.007
Method:                Least Squares   F-statistic:                     623.7
Date:               Sun, 04 Jun 2023   Prob (F-statistic):           3.34e-137
```

```
Time:                       09:35:30   Log-Likelihood:            -3.0107e+05
No. Observations:              92425   AIC:                         6.021e+05
Df Residuals:                  92423   BIC:                         6.022e+05
Df Model:                          1
Covariance Type:           nonrobust
================================================================================
                 coef    std err          t      P>|t|      [0.025      0.975]
--------------------------------------------------------------------------------
Intercept      3.2188      0.047     68.142      0.000       3.126       3.311
ydstogo        0.1329      0.005     24.974      0.000       0.122       0.143
================================================================================
Omnibus:                   81985.726   Durbin-Watson:                    1.994
Prob(Omnibus):                 0.000   Jarque-Bera (JB):          4086040.920
Skew:                          4.126   Prob(JB):                          0.00
Kurtosis:                     34.511   Cond. No.                          20.5
================================================================================
```

Notes:
[1] Standard Errors assume that the covariance matrix of the errors is
    correctly specified.

This summary output includes a description of the model. Many of the summary items should be straightforward, such as the dependent variable (Dep. Variable), Date, and Time. The number of observations (No. Observations) relates to the degrees of freedom. The *degrees of freedom* indicate the number of "extra" observations that exist compared to the number of parameters fit.

With this model, two parameters were fit: a slope for rushing_yards and an intercept, Intercept. Hence, the Df Residuals equals the No. Observations–2. The R^2 value corresponds to how well the model fits the data. If $R^2 = 1.0$, the model fits the data perfectly. Conversely, if $R^2 = 0$, the model does not predict the data at all. In this case, the low R^2 of 0.007 shows that the simple model does not predict the data well.

Other outputs of interest include the coefficient estimates for the Intercept and ydstogo. The Intercept is the number of rushing yards expected to be gained if there are 0 yards to go for a first down or a touchdown (which never actually occurs in real life). The slope for ydstogo corresponds to the expected number of additional rushing yards expected to be gained for each additional yard to go. For example, a rushing play with 2 yards to go would be expected to produce 3.2(intercept) + 0.1(slope) × 2(number of yards to go) = 3.4 yards on average.

With the coefficients, a point estimate exists (coef) as well as the standard error (std err). The *standard error (SE)* captures the uncertainty around the estimate for the coefficient, something Appendix B describes in greater detail. The t-value comes from a statistical distribution (specifically, the *t*-distribution) and is used to generate the SE and confidence interval (CI). The *p*-value provides the probability of obtaining the observed *t*-value, assuming the null hypothesis of the coefficient being 0 is true.

The *p*-value ties into null hypothesis significance testing (NHST), something that most introductory statistics courses cover, but is increasingly falling out of use by practicing statisticians. Lastly, the summary includes the 95% CI for the coefficients. The lower CI is the [0.025 column, and the upper CI is the 0.975] column (97.5 – 2.5 = 95%). The *95% CI* should contain the true estimate for the coefficient 95% of the time, assuming the observation process is repeated many times. However, you never know *which* 5% of the time you are wrong.

You can fit a similar, linear model (lm()) by using R and then print the summary results:

```
## R
yard_to_go_r <-
    lm(rushing_yards ~ 1 + ydstogo, data = pbp_r_run)

summary(yard_to_go_r)
```

Resulting in:

```
Call:
lm(formula = rushing_yards ~ 1 + ydstogo, data = pbp_r_run)

Residuals:
    Min      1Q  Median      3Q     Max
-33.079  -3.352  -1.415   1.453  94.453

Coefficients:
             Estimate Std. Error t value Pr(>|t|)
(Intercept)  3.21876    0.04724   68.14   <2e-16 ***
ydstogo      0.13287    0.00532   24.97   <2e-16 ***
---
Signif. codes:  0 '***' 0.001 '**' 0.01 '*' 0.05 '.' 0.1 ' ' 1

Residual standard error: 6.287 on 92423 degrees of freedom
Multiple R-squared:  0.006703,  Adjusted R-squared:  0.006692
F-statistic: 623.7 on 1 and 92423 DF,  p-value: < 2.2e-16
```

The general structure of the regression output differs in R compared to Python. However, the items provided are similar, with the main difference occurring in formatting. R provides the model formula, or Call, followed by a summary of the residuals. *Residuals* indicate how well data compares to the model's fit. For RYOE, you actually will use the residuals later. Then the summary provides the Coefficients and their uncertainty. Last, model details are printed, which are similar to the Python details.

Checking degrees of freedom may seem strange to people starting out modeling. However, this can be a great check for your data to make sure the model is using all your inputs correctly and that values are not being lost. One of our friends, Barb Bennie, spends most of a semester teaching her graduate students in statistics how to compare degrees of freedom across models. When writing this book, the Python and R versions of the nflfastR data were giving different values for model estimates, and the degrees of freedom helped us figure out the packages needed to be updated on our machines. Do not underestimate the utility and power of understanding degrees of freedom.

Lastly, before moving on to look at RYOE, we need to save the residuals to create an RYOE column in the data. *Residuals* are the difference between a model's expected (or predicted) output and the observed data. With pandas in Python, create a new RYOE column in the pbp_py_run dataframe from the model's residuals:

```Python
## Python
pbp_py_run["ryoe"] =\
    yard_to_go_py\
    .fit()\
    .resid
```

 Linear models in Python and R have capabilities and tools that we only scratch the surface of in this book. Learning the details of these tools, using resources such as those listed in "Suggested Readings" on page 77 will help you better unlock the power of linear models.

Likewise, in R mutate the pbp_r_run data to create a new column, ryoe:

```R
## R
pbp_r_run <-
    pbp_r_run |>
    mutate(ryoe = resid(yard_to_go_r))
```

R, which was created for teaching statistics, is based on the S language. Given this history and the state of statistics in the early 1990s, R has linear models well integrated into the language. In contrast, Python has a clone of R for linear models for statistical inference—specifically, the `statsmodels` package. The main package for models in Python, `scikit-learn` (`sklearn`) focuses on machine learning rather than statistical inference. Understanding the history of R and Python can provide insight into *why* the languages exist as they do as well as enable you to leverage their respective strengths. We would also argue that if all you need and want to do is fit regression models for statistical inference, R would be the better software choice.

Who Was the Best in RYOE?

Now, look at the leaderboard for RYOE from 2016 to 2022, first in total yards over expected and average yards over expected per carry. As with the passer data in Chapter 2, you will need to group by both the `rusher` and `rusher_id` because some players have the same last name and first initial.

With `pandas`, group by `seasons`, `rusher_id`, and `rusher`. Then, aggregate `ryoe` with the count, sum, and mean, and aggregate `rushing_yards` with the mean. This gives the following columns:

- The `count` of RYOE is the *number of carries a rusher has*.
- The `sum` of RYOE is the *total RYOE*.
- The `mean` of RYOE is the *RYOE per carry*.
- The `mean` of rushing yards is the *yards per carry*.

Flatten the columns and reset the index to make the dataframe easier to work with. Next, rename the columns to give them football-specific names. Lastly, query the result to print only players with more than 50 carries:

```Python
## Python
ryoe_py =\
    pbp_py_run\
    .groupby(["season", "rusher_id", "rusher"])\
    .agg({
        "ryoe": ["count", "sum", "mean"],
        "rushing_yards": "mean"})

ryoe_py.columns = \
    list(map("_".join, ryoe_py.columns))
ryoe_py.reset_index(inplace=True)

ryoe_py =\
```

```
ryoe_py\
    .rename(columns={
        "ryoe_count": "n",
        "ryoe_sum": "ryoe_total",
        "ryoe_mean": "ryoe_per",
        "rushing_yards_mean": "yards_per_carry",
    }
).query("n > 50")

print(ryoe_py.sort_values("ryoe_total", ascending=False))
```

Resulting in:

```
      season  rusher_id     rusher    n  ryoe_total  ryoe_per  yards_per_carry
1989    2021  00-0036223   J.Taylor  332  417.501295  1.257534         5.454819
1440    2020  00-0032764    D.Henry  397  362.768406  0.913774         5.206549
1258    2019  00-0034796  L.Jackson  135  353.652105  2.619645         6.800000
1143    2019  00-0032764    D.Henry  387  323.921354  0.837006         5.131783
1474    2020  00-0033293    A.Jones  222  288.358241  1.298911         5.540541
...      ...         ...        ...  ...         ...       ...              ...
419     2017  00-0029613   D.Martin  139 -198.461432 -1.427780         2.920863
122     2016  00-0029613   D.Martin  144 -199.156646 -1.383032         2.923611
675     2018  00-0027325   L.Blount  155 -247.528360 -1.596957         2.696774
1058    2019  00-0030496     L.Bell  245 -286.996618 -1.171415         3.220408
267     2016  00-0032241   T.Gurley  278 -319.803875 -1.150374         3.183453

[534 rows x 7 columns]
```

To print the entire table in Python once, run `print(ryoe_py.query("n > 50").to_string())`, something we did not do in order to save space. Alternatively, you can change the printing in your entire session by using the `pandas set_option()` function. For example, `pd.set_option("display.min_rows", 10)` would always print 10 rows.

With R, use the `pbp_r_run` data and group by `season`, `rusher_id`, and `rusher`. Then `summarize()` to get the number per group, total RYOE, average RYOE, and yards per carry. Lastly, filter to include only players with more than 50 carries:

```
## R
ryoe_r <-
    pbp_r_run |>
    group_by(season, rusher_id, rusher) |>
    summarize(
        n = n(),
        ryoe_total = sum(ryoe),
        ryoe_per = mean(ryoe),
        yards_per_carry = mean(rushing_yards)
    ) |>
    arrange(-ryoe_total) |>
    filter(n > 50)

print(ryoe_r)
```

Resulting in:

```
# A tibble: 534 × 7
# Groups:   season, rusher_id [534]
   season rusher_id  rusher          n ryoe_total ryoe_per yards_per_carry
    <dbl> <chr>      <chr>       <int>      <dbl>    <dbl>           <dbl>
 1   2021 00-0036223 J.Taylor      332       418.    1.26             5.45
 2   2020 00-0032764 D.Henry       397       363.    0.914            5.21
 3   2019 00-0034796 L.Jackson     135       354.    2.62             6.8
 4   2019 00-0032764 D.Henry       387       324.    0.837            5.13
 5   2020 00-0033293 A.Jones       222       288.    1.30             5.54
 6   2019 00-0031687 R.Mostert     190       282.    1.48             5.83
 7   2016 00-0033045 E.Elliott     344       279.    0.810            5.10
 8   2021 00-0034791 N.Chubb       228       276.    1.21             5.52
 9   2022 00-0034796 L.Jackson      73       276.    3.78             7.82
10   2020 00-0034791 N.Chubb       221       254.    1.15             5.48
# i 524 more rows
```

We did not have you print out all the tables, to save space. However, using |>
print(n = Inf) at the end would allow you to see the entire table in R. Alternatively, you could change all printing for an R session by running options(pil
lar.print_min = n).

For the filtered lists, we had you print only the lists with players who carried the ball
50 or more times to leave out outliers as well as to save page space. We don't have to
do that for total yards, since players with so few carries will not accumulate that many
RYOE, anyway.

By total RYOE, 2021 Jonathan Taylor was the best running back in football since
2016, generating over 400 RYOE, follow by the aforementioned Henry, who generated 374 RYOE during his 2,000-yard 2020 season. The third player on the list, Lamar
Jackson, is a quarterback, who in 2019 earned the NFL's MVP award for both rushing
for over 1,200 yards (an NFL record for a quarterback) and leading the league in
touchdown passes. In April of 2022, Jackson signed the richest contract in NFL
history for his efforts.

One interesting quirk of the NFL data is that only *designed runs* for quarterbacks
(plays that are actually running plays and not broken-down passing plays, where
the quarterback pulls the ball down and runs with it) are counted in this dataset.
So Jackson generating this much RYOE on just a subset of his runs is incredibly
impressive.

Next, sort the data by RYOE per carry. We include only code for R, but the previous
Python code is readily adaptable:

```
## R
ryoe_r |>
    arrange(-ryoe_per)
```

Resulting in:

```
# A tibble: 534 × 7
# Groups:   season, rusher_id [534]
   season rusher_id   rusher           n ryoe_total ryoe_per yards_per_carry
    <dbl> <chr>       <chr>        <int>      <dbl>    <dbl>           <dbl>
 1   2022 00-0034796 L.Jackson       73       276.     3.78            7.82
 2   2019 00-0034796 L.Jackson      135       354.     2.62            6.8
 3   2019 00-0035228 K.Murray        56       122.     2.17            6.5
 4   2020 00-0034796 L.Jackson      121       249.     2.06            6.26
 5   2021 00-0034750 R.Penny        119       229.     1.93            6.29
 6   2022 00-0036945 J.Fields        85       160.     1.88            6
 7   2022 00-0033357 T.Hill          96       178.     1.86            5.99
 8   2021 00-0034253 D.Hilliard      56       101.     1.80            6.25
 9   2022 00-0034750 R.Penny         57        99.2    1.74            6.07
10   2019 00-0034400 J.Wilkins       51        87.8    1.72            6.02
# i 524 more rows
```

Looking at RYOE per carry yields three Jackson years in the top four, sandwiching a year by Kyler Murray, who is also a quarterback. Murray was the first-overall pick in the 2019 NFL Draft and elevated an Arizona Cardinals offense from the worst in football in 2018 to a much more respectable standing in 2019, through a combination of running and passing. Rashaad Penny, the Seattle Seahawks' first-round pick in 2018, finally emerged in 2021 to earn almost 2 yards over expected per carry while leading the NFL in yards per carry overall (6.3). This earned Penny a second contract with Seattle the following offseason.

A reasonable question we should ask is whether total yards or yards per carry is a better measure of a player's ability. Fantasy football analysts and draft analysts both carry a general consensus that "volume is earned." The idea is that a hidden signal is in the data when a player plays enough to generate a lot of carries.

Data is an incomplete representation of reality, and players do things that aren't captured. If a coach plays a player a lot, it's a good indication that that player is good. Furthermore, a negative relationship generally exists between volume and efficiency for that same reason. If a player is good enough to play a lot, the defense is able to key on him more easily and reduce his efficiency. This is why we often see backup running backs with high yards-per-carry values relative to the starter in front of them (for example, look at Tony Pollard and Ezekiel Elliott for the Dallas Cowboys in 2019–2022). Other factors are at play as well (such as Zeke's draft status and contract) that don't necessarily make volume the be-all and end-all, but it's important to inspect.

Is RYOE a Better Metric?

Anytime you create a new metric for player or team evaluation in football, you have to test its predictive power. You can put as much thought into your new metric as the next person, and the adjustments, as in this chapter, can be well-founded. But if the metric is not more stable than previous iterations of the evaluation, you either have to conclude that the work was in vain or that the underlying context that surrounds a player's performance is actually the entity that carries the signal. Thus, you're attributing too much of what is happening in terms of production on the individual player.

In Chapter 2, you conducted stability analysis on passing data. Now, you will look at RYOE per carry for players with 50 or more carries versus traditional yards-per-carry values. We don't have you look at total RYOE or rushing yards since that embeds two measures of performance (volume and efficiency). Volume is in both measures, which muddies things.

This code is similar to that from "Deep Passes Versus Short Passes" on page 41, and we do not provide a walk-through here, simply the code with comments. Hence, we refer you to that section for details.

In Python, use this code:

```python
## Python
#  keep only columns needed
cols_keep =\
    ["season", "rusher_id", "rusher",
     "ryoe_per", "yards_per_carry"]

# create current dataframe
ryoe_now_py =\
    ryoe_py[cols_keep].copy()

# create last-year's dataframe
ryoe_last_py =\
    ryoe_py[cols_keep].copy()

# rename columns
ryoe_last_py\
    .rename(columns = {'ryoe_per': 'ryoe_per_last',
                       'yards_per_carry': 'yards_per_carry_last'},
            inplace=True)

# add 1 to season
ryoe_last_py["season"] += 1

# merge together
ryoe_lag_py =\
    ryoe_now_py\
    .merge(ryoe_last_py,
```

```
                 how='inner',
                 on=['rusher_id', 'rusher',
                     'season'])
```

Lastly, examine the correlation for yards per carry:

```
## Python
ryoe_lag_py[["yards_per_carry_last", "yards_per_carry"]].corr()
```

Resulting in:

	yards_per_carry_last	yards_per_carry
yards_per_carry_last	1.00000	0.32261
yards_per_carry	0.32261	1.00000

Repeat with RYOE:

```
## Python
ryoe_lag_py[["ryoe_per_last", "ryoe_per"]].corr()
```

Which results in:

	ryoe_per_last	ryoe_per
ryoe_per_last	1.000000	0.348923
ryoe_per	0.348923	1.000000

With R, use this code:

```
## R
# create current dataframe
ryoe_now_r <-
    ryoe_r |>
    select(-n, -ryoe_total)

# create last-year's dataframe
# and add 1 to season
ryoe_last_r <-
    ryoe_r |>
    select(-n, -ryoe_total) |>
    mutate(season = season + 1) |>
    rename(ryoe_per_last = ryoe_per,
           yards_per_carry_last = yards_per_carry)

# merge together
ryoe_lag_r <-
    ryoe_now_r |>
    inner_join(ryoe_last_r,
               by = c("rusher_id", "rusher", "season")) |>
    ungroup()
```

Then select the two yards-per-carry columns and examine the correlation:

```
## R
ryoe_lag_r |>
    select(yards_per_carry, yards_per_carry_last) |>
    cor(use = "complete.obs")
```

Resulting in:

```
                   yards_per_carry yards_per_carry_last
yards_per_carry          1.0000000            0.3226097
yards_per_carry_last     0.3226097            1.0000000
```

Repeat the correlation with the RYOE columns:

```
## R
ryoe_lag_r |>
    select(ryoe_per, ryoe_per_last) |>
    cor(use = "complete.obs")
```

Resulting in:

```
              ryoe_per ryoe_per_last
ryoe_per     1.0000000     0.3489235
ryoe_per_last 0.3489235    1.0000000
```

These results show that, for players with more than 50 rushing attempts in back-to-back seasons, this version of RYOE per carry is slightly more stable year to year than yards per carry (because the correlation coefficient is larger). Therefore, yards per carry includes information inherent to the specific play in which a running back carries the ball. Furthermore, this information can vary from year to year. After extracted out, our new metric for running back performance is slightly more predictive year to year.

As far as the question of whether (or, more accurately, how much) running backs matter, the difference between our correlation coefficients suggest that we don't have much in the way of a conclusion to be drawn so far. Prior to this chapter, you've really looked at statistics for only one position (quarterback), and the more stable metric in the profile for that position (yards per pass attempt on short passes, with r values nearing 0.5) is much more stable than yards per carry and RYOE. Furthermore, you haven't done a thorough analysis of the running game relative to the passing game, which is essential to round out any argument that running backs are or are not important.

The key questions for football analysts about running backs are as follows:

1. Are running back contributions valuable?
2. Are their contributions repeatable across years?

In Chapter 4, you will add variables to control for other factors affecting the running game. For example, the *down* part of down and distance is surely important, because a defense will play even tighter on fourth down and 1 yard to go than it will on third down and that same 1 yard to go. A team trailing (or losing) by 14 points will have an easier time of running the ball than a team up by 14 points (all else being equal) because a team that is ahead might be playing "prevent" defense, a defense that is willing to allow yards to their opponent, just not *too* many yards.

Data Science Tools Used in This Chapter

This chapter covered the following topics:

- Fitting a simple linear regression by using `OLS()` in Python or by using `lm()` in R
- Understanding and reading the coefficients from a simple linear regression
- Plotting a simple linear regression by using `regplot()` from `seaborn` in Python or by using `geom_smooth()` in R
- Using correlations with the `corr()` function in Python and R to conduct stability analysis

Exercises

1. What happens if you repeat the correlation analysis with 100 carries as the threshold? What happens to the differences in *r* values?

2. Assume all of Alstott's carries were on third down and 1 yard to go, while all of Dunn's carries came on first down and 10 yards to go. Is that enough to explain the discrepancy in their yards-per-carry values (3.7 versus 4.0)? Use the coefficient from the simple linear model in this chapter to understand this question.

3. What happens if you repeat the analyses in this chapter with yards to go to the endzone (`yardline_100`) as your feature?

4. Repeat the processes within this chapter with receivers and the passing game. To do this, you have to filter by `play_type == "pass"` and `receiver_id` not being NA or NULL.

Suggested Readings

Eric wrote several articles on running backs and running back value while at PFF. Examples include the following:

- "The NFL's Best Running Backs on Perfectly and Non-perfectly Blocked Runs in 2021" (*https://oreil.ly/IArDE*)
- "Are NFL Running Backs Easily Replaceable: The Story of the 2018 NFL Season" (*https://oreil.ly/x5dAk*)
- "Explaining Dallas Cowboys RB Ezekiel Elliott's 2018 PFF Grade" (*https://oreil.ly/aYhfj*)

Additionally, many books exist on regression and introductory statistics and "Suggested Readings" on page 285 lists several introductory statistics books. For regression, here are some books we found useful:

- *Regression and Other Stories* by Andrew Gelman et al. (Cambridge University Press, 2020). This book shows how to apply regression analysis to real-world problems. For those of you looking for more worked case studies, we recommend this book to help you learn how to think about applying regression.
- *Regression Modeling Strategies: With Applications to Linear Models, Logistic and Ordinal Regression, and Survival Analysis*, 2nd edition, by Frank E. Harrell Jr. (Springer, 2015). This book helped one of the authors think through the world of regression modeling. It is advanced but provides a good oversight into regression analysis. The book is written at an advanced undergraduate or introductory graduate level. Although hard, working through this book provides mastery of regression analysis.

## CHAPTER 4

Multiple Regression: Rushing Yards Over Expected

In the previous chapter, you looked at rushing yards through a different lens, by controlling for the number of yards needed to gain for a first down or a touchdown. You found that common sense prevails; the more yards a team needs to earn for a first down or a touchdown, the easier it is to gain yards on the ground. This tells you that adjusting for such things is going to be an important part of understanding running back play.

A limitation of simple linear regression is that you will have more than one important variable to adjust for in the running game. While distance to first down or touchdown matters a great deal, perhaps the down is important as well. A team is more likely to run the ball on first down and 10 yards to go than it is to run the ball on third down and 10 yards to go, and hence the defense is more likely to be more geared up to stop the run on the former than the latter.

Another example is point differential. The score of the game influences the expectation in multiple ways, as a team that is ahead by a lot of points will not crowd the line of scrimmage as much as a team that is locked into a close game with its opponent. In general, a plethora of variables need to be *controlled for* when evaluating a football play. The way we do this is through multiple linear regression.

Definition of Multiple Linear Regression

We know that running the football isn't affected by just one thing, so we need to build a model that predicts rushing yards, but includes more features to account for other things that might affect the prediction. Hence, to build upon Chapter 3, we need multiple regression.

Multiple regression estimates for the effect of several (*multiple*) predictors on a single response by using a linear combination (or *regression*) of the predictor variables. Chapter 3 presented *simple linear regression*, which is a special case of multiple regression. In a simple linear regression, two parameters exist: an *intercept* (or average value) and a *slope*. These model the effect of a continuous predictor on the response. In Chapter 3, the Python/R formula for your simple linear regression had rushing yards predicted by an intercept and yards to go: `rushing_yards ~ 1 + ydstogo`.

However, you may be interested in multiple predictor variables. For example, consider the multiple regression that *corrects* for down when estimating expected rushing yards. You would use an equation (or formula) that predicts rushing yards based on yards to go and down: `rushing_yards ~ ydstogo + down`. Yards to go is an example of a *continuous* predictor variable. Informally, think of *continuous predictors* as numbers such as a players' weights, like 135, 302, or 274. People sometimes use the term *slope* for continuous predictor variables. Down is an example of a *discrete* predictor variable. Informally, think of discrete predictors as groups or categories such as position (like running back or quarterback). By default, the `formula` option in `statsmodels` and base R treat discrete predictors as *contrasts*.

Let's dive into the example formula, `rushing_yards ~ ydstogo + down`. R would estimate an intercept for the lowest (or alphabetically first) down and then the *contrast*, or difference, for the other downs. For example, four predictors would be estimated: an intercept that is the mean `rushing_yards` for yards to go, a contrast for second downs compared to first down, a contrast for third downs compared to first down, and a contrast for fourth downs compared to first down. To see this, look at the *design matrix*, or *model matrix*, in R (Python has similar features that are not shown here). First, create a demonstration dataset:

```R
## R
library(tidyverse)

demo_data_r <- tibble(down = c("first", "second"),
                      ydstogo = c(10, 5))
```

Then, create a model matrix by using the formula's righthand side and the `model.matrix()` function:

```R
## R
model.matrix(~ ydstogo + down,
             data = demo_data_r)
```

Resulting in:

```
  (Intercept) ydstogo downsecond
1           1      10          0
2           1       5          1
attr(,"assign")
[1] 0 1 2
```

```
attr(,"contrasts")
attr(,"contrasts")$down
[1] "contr.treatment"
```

Notice that the output has three columns: an intercept, a slope for `ydstogo`, and a contrast for second down (`downsecond`).

However, you can also estimate an intercept for each down by using a `-1`, such as `rushing_yards ~ ydstogo + down -1`. This would estimate four predictors: an intercept for first down, an intercept for second down, an intercept for third down, and an intercept for fourth down. Use R to look at the example model matrix:

```
## R
model.matrix(~ ydstogo + down - 1,
             data = demo_data_r)
```

Resulting in:

```
  ydstogo downfirst downsecond
1      10         1          0
2       5         0          1
attr(,"assign")
[1] 1 2 2
attr(,"contrasts")
attr(,"contrasts")$down
[1] "contr.treatment"
```

Notice that you have the same number of columns as before, but each down has its own column.

 Computer languages such as Python and R get confused by some groups such as down. The computer tries to treat these predictors as continuous, such as the numbers 1, 2, 3, and 4 rather than first down, second down, third down, and fourth down. In pandas you will change `down` to be a string (`str`), and in R, you will change `down` to be a character.

The formula for multiple regression allows for many discrete and continuous predictors. When multiple discrete predictors are present, such as `down + team`, the first variable (`down`, in this case) can either be estimated as an intercept or contrast parameters. All other discrete predictors are estimated as contrasts, with the first groupings treated as part of the intercept. Rather than getting caught up in thinking about slopes and intercepts, you can use the term *coefficients* to describe the estimated predictor variables for multiple regression.

Exploratory Data Analysis

In the case of rushing yards, you're going to use the following variables as *features* in the multiple linear regression model: down (down), distance (ydstogo), yards to go to the endzone (yardline_100), run location (run_location), and score differential (score_differential). Other variables could also be used, of course, but for now you're using these variables in large part because they all affect rushing yards in one way or another.

First, load in the data and packages you will be using, filter for only the run data (as you did in Chapter 3), and remove plays that were not a regular down. Do this with Python:

```python
import pandas as pd
import numpy as np
import nfl_data_py as nfl
import statsmodels.formula.api as smf
import matplotlib.pyplot as plt
import seaborn as sns

seasons = range(2016, 2022 + 1)
pbp_py = nfl.import_pbp_data(seasons)

pbp_py_run = \
    pbp_py\
    .query('play_type == "run" & rusher_id.notnull() &' +
           "down.notnull() & run_location.notnull()")\
    .reset_index()

pbp_py_run\
    .loc[pbp_py_run.rushing_yards.isnull(), "rushing_yards"] = 0
```

Or with R:

```r
library(tidyverse)
library(nflfastR)

pbp_r <- load_pbp(2016:2022)

pbp_r_run <-
    pbp_r |>
    filter(play_type == "run" & !is.na(rusher_id) &
        !is.na(down) & !is.na(run_location)) |>
    mutate(rushing_yards = ifelse(is.na(rushing_yards),
        0,
        rushing_yards
    ))
```

Next, let's create a histogram for down and rushing yards gained with Python in Figure 4-1:

```python
## Python
# Change theme for chapter
sns.set_theme(style="whitegrid", palette="colorblind")

# Change down to be an integer
pbp_py_run.down =\
    pbp_py_run.down.astype(str)

# Plot rushing yards by down
g = \
    sns.FacetGrid(data=pbp_py_run,
                  col="down", col_wrap=2);
g.map_dataframe(sns.histplot, x="rushing_yards");
plt.show();
```

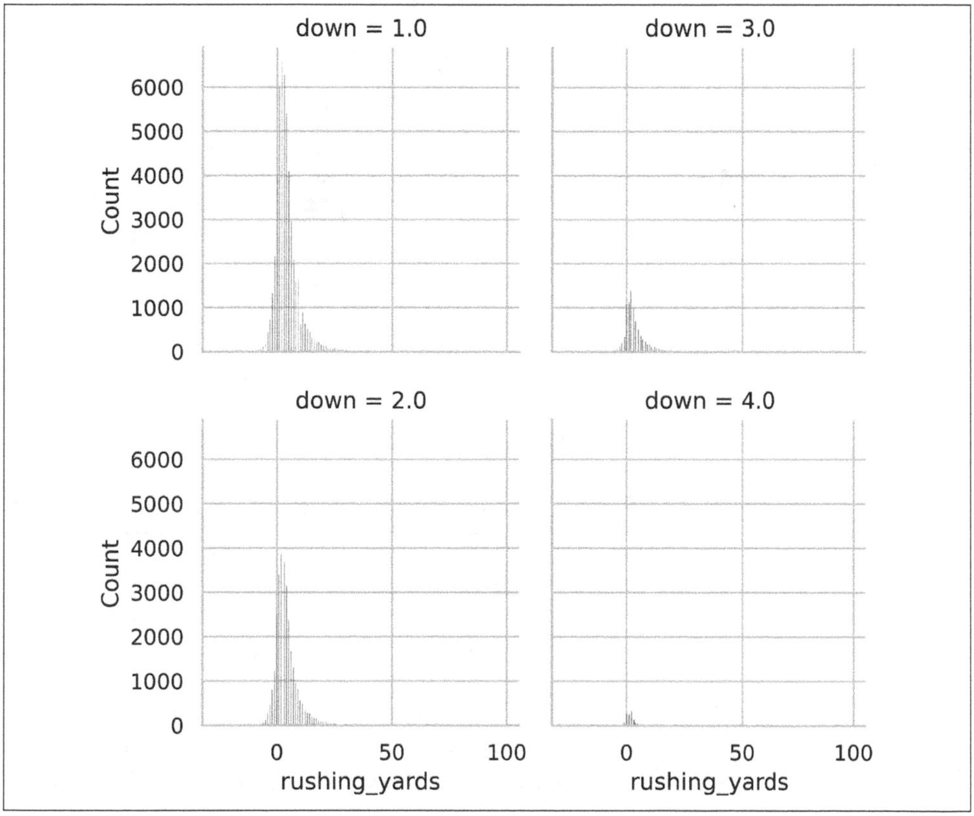

Figure 4-1. Histogram of rushing yards by downs with *seaborn*

Or, Figure 4-2 with R:

```R
## R
# Change down to be an integer
pbp_r_run <-
    pbp_r_run |>
    mutate(down = as.character(down))

# Plot rushing yards by down
ggplot(pbp_r_run, aes(x = rushing_yards)) +
    geom_histogram(binwidth = 1) +
    facet_wrap(vars(down), ncol = 2,
               labeller = label_both) +
    theme_bw() +
    theme(strip.background = element_blank())
```

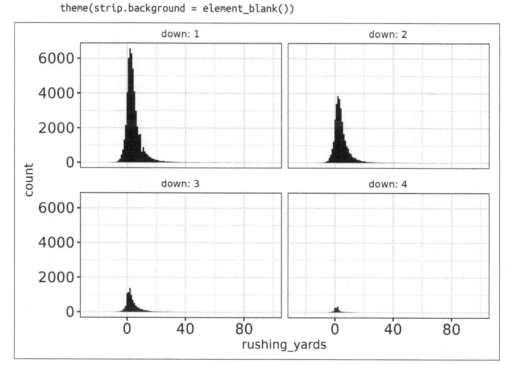

Figure 4-2. Histogram of rushing yards by downs with `ggplot2`

This is interesting, as it looks like down decreases rushing yards. However, the data has a confounder, which is that rushes often happen on late downs with smaller distances. Let's look at only situations where `ydstogo == 10`. In Python, create Figure 4-3:

```
## Python
sns.boxplot(data=pbp_py_run.query("ydstogo == 10"),
            x="down",
            y="rushing_yards");
plt.show()
```

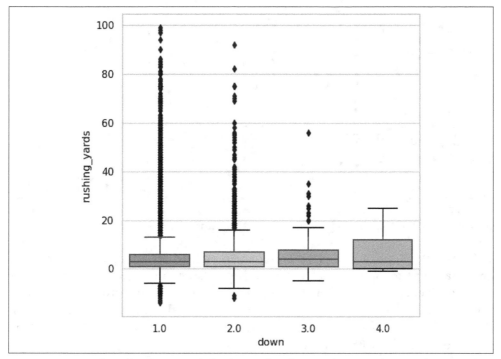

Figure 4-3. Boxplot of rushing yards by downs for plays with 10 yards to go (seaborn)

Or with R, create Figure 4-4:

```
## R
pbp_r_run |>
    filter(ydstogo == 10) |>
    ggplot(aes(x = down, y = rushing_yards)) +
    geom_boxplot() +
    theme_bw()
```

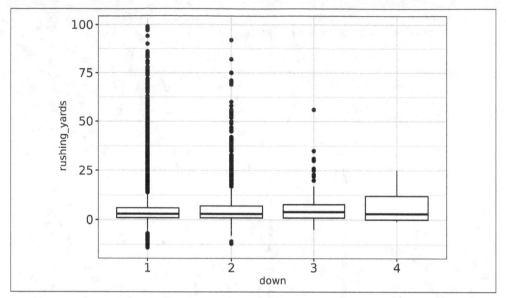

Figure 4-4. Boxplot of rushing yards by downs for plays with 10 yards to go (ggplot2)

OK, now you see what you expect. This is an example of *Simpson's paradox*: including an extra, third grouping variable changes the relationship between the two other variables. Nonetheless, it's clear that down affects the rushing yards on a play and should be accounted for. Similarly, let's look at yards to the endzone in seaborn with Figure 4-5 (and change the transparency with scatter_kws={'alpha':0.25} and the regression line color with line_kws={'color': 'red'}):

```Python
## Python
sns.regplot(
    data=pbp_py_run,
    x="yardline_100",
    y="rushing_yards",
    scatter_kws={"alpha": 0.25},
    line_kws={"color": "red"}
);
plt.show();
```

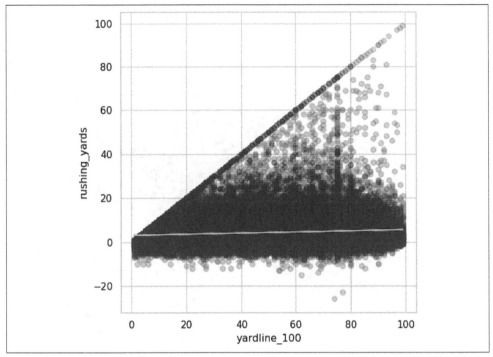

*Figure 4-5. Scatterplot with linear trendline for ball position (yards to go to the endzone) and rushing yards from a play (*seaborn*)*

Or with R, create Figure 4-6 (and change the transparency with `alpha = 0.25` to help you see the overlapping points):

```
## R
ggplot(pbp_r_run, aes(x = yardline_100, y = rushing_yards)) +
    geom_point(alpha = 0.25) +
    stat_smooth(method = "lm") +
    theme_bw()
```

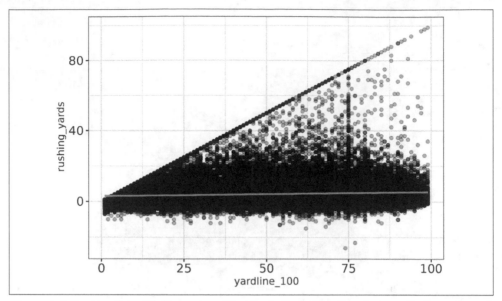

Figure 4-6. Scatterplot with linear trendline for ball position (yards to go to the endzone) and rushing yards from a play (ggplot2)

This doesn't appear to do much, but let's look at what happens after you bin and average with Python:

```python
## Python
pbp_py_run_y100 =\
    pbp_py_run\
    .groupby("yardline_100")\
    .agg({"rushing_yards": ["mean"]})

pbp_py_run_y100.columns =\
    list(map("_".join, pbp_py_run_y100.columns))

pbp_py_run_y100.reset_index(inplace=True)
```

Now, use these results to create Figure 4-7:

```
sns.regplot(
    data=pbp_py_run_y100,
    x="yardline_100",
    y="rushing_yards_mean",
    scatter_kws={"alpha": 0.25},
    line_kws={"color": "red"}
);
plt.show();
```

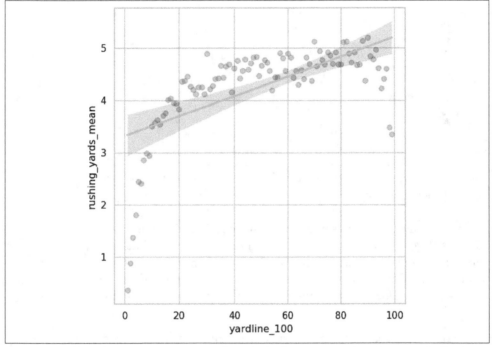

Figure 4-7. Scatterplot with linear trendline for ball position and rushing yards for data binned by yard (seaborn)

Or with R, create Figure 4-8:

```
## R
pbp_r_run |>
    group_by(yardline_100) |>
    summarize(rushing_yards_mean = mean(rushing_yards)) |>
    ggplot(aes(x = yardline_100, y = rushing_yards_mean)) +
    geom_point() +
    stat_smooth(method = "lm") +
    theme_bw()
```

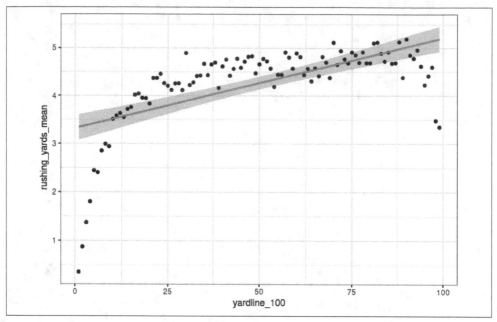

Figure 4-8. Scatterplot with linear trendline for ball position and rushing yards for data binned by yard (`ggplot2`)

Figures 4-7 and 4-8 show some football insight. Running plays with less than about 15 yards to go are limited by distance because there is limited distance to the endzone and tougher red-zone defense. Likewise, plays with more than 90 yards to go take the team out of its own end zone. So, defense will be trying hard to force a safety, and offense will be more likely to either punt or play conservatively to avoid allowing a safety.

Here you get a clear positive (but nonlinear) relationship between average rushing yards and yards to go to the endzone, so it benefits you to include this feature in the models. In "Assumption of Linearity" on page 108, you can see what happens to the model if values less than 15 yards or greater than 90 yards are removed. In practice, more complicated models can effectively deal with these nonlinearities, but we save that for a different book. Now, you look at run location with Python in Figure 4-9:

```
## Python
sns.boxplot(data=pbp_py_run,
            x="run_location",
            y="rushing_yards");
plt.show();
```

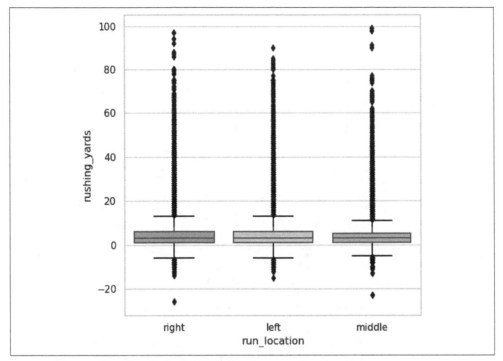

Figure 4-9. Boxplot of rushing yards by run location (seaborn)

Or with R in Figure 4-10:

```
## R
ggplot(pbp_r_run, aes(run_location, rushing_yards)) +
    geom_boxplot() +
    theme_bw()
```

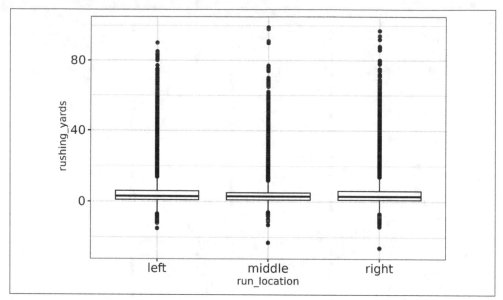

Figure 4-10. Boxplot of rushing yards by run location (ggplot2)

Not only are the means/medians slightly different here, but the variances/interquartile ranges also appear to vary, so keep them in the models. Another comment about the run location is that the 75th percentile in each case is really low. Three-quarters of the time, regardless of whether a player goes left, right, or down the middle, he goes 10 yards or less. Only in a few rare cases do you see a long rush.

Lastly, look at score differential, using the binning and aggregating you used for yards to go to the endzone in Python:

```python
## Python
pbp_py_run_sd = \
    pbp_py_run\
    .groupby("score_differential")\
    .agg({"rushing_yards": ["mean"]}
)

pbp_py_run_sd.columns =\
    list(map("_".join, pbp_py_run_sd.columns))

pbp_py_run_sd.reset_index(inplace=True)
```

Now, use these results to create Figure 4-11:

```python
## Python
sns.regplot(
    data=pbp_py_run_sd,
    x="score_differential",
    y="rushing_yards_mean",
    scatter_kws={"alpha": 0.25},
    line_kws={"color": "red"}
);
plt.show();
```

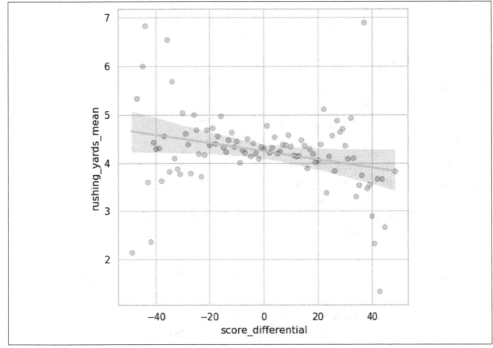

Figure 4-11. Scatterplot with linear trendline for score differential and rushing yards, for data binned by score differential (seaborn)

Or in R, create Figure 4-12:

```r
## R
pbp_r_run |>
    group_by(score_differential) |>
    summarize(rushing_yards_mean = mean(rushing_yards)) |>
    ggplot(aes(score_differential, rushing_yards_mean)) +
    geom_point() +
    stat_smooth(method = "lm") +
    theme_bw()
```

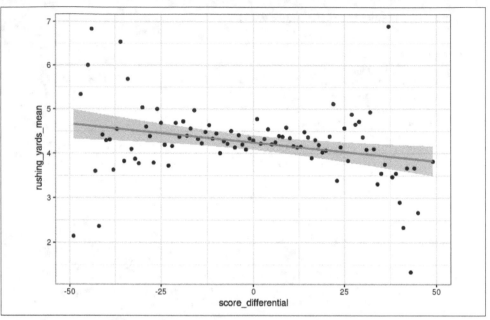

Figure 4-12. Scatterplot with linear trendline for score differential and rushing yards, for data binned by score differential (`ggplot2`)

You can see a clear negative relationship, as hypothesized previously. Hence you will leave score differential in the model.

> When you see code for plots in this book, think about how you would want to improve the plots. Also, look at the code, and if you do not understand some arguments, search and figure out how to change the plotting code. The best way to become a better data plotter is to explore and create your own plots.

Applying Multiple Linear Regression

Now, apply the multiple linear regression to rederive rushing yards over expected (RYOE). In this section, we gloss over steps that are covered in Chapter 3.

First, fit the model. Then save the calculated residuals as the RYOE. Recall that residuals are the difference between the value predicted by the model and the value observed in the data. This could be calculated directly by taking the observed rushing yards and subtracting the predicted rushing yards from the model. However, residuals are commonly used in statistics, and Python and R both include residuals as part of the model's fit. This derivation differs from the method in Chapter 3 because you have created a more complicated model.

The model predicts `rushing_yards` by creating the following:

- An intercept (1)
- A term contrasting the second, third, and fourth downs to the first down (down)
- A coefficient for `ydstogo`
- An *interaction* between `ydstogo` and `down` that estimates a `ydstogo` contrast for each `down` (`ydstogo:down`)
- A coefficient for yards to go to the endzone (`yardline_100`)
- The location of the running play on the field (`run_location`)
- The difference between each team's scores (`score_differential`)

 Multiple approaches exist for using formulas to indicate interactions. The longest, but most straightforward approach for our example would be `down + ydstogo + as.factor(down):ydstogo`. This may be abbreviated as `down * ydstogo`. Thus, the example formula, `rushing_yards ~ 1 + down + ydstogo + down:ydstogo + yardline_100 + run_location + score_differential`, could be written as `rushing_yards ~ down * ydstogo + yardline_100 + run_location + score_differential` and saves writing three terms.

In Python, fit the model with the `statsmodels` package; then save the residuals as RYOE:

```Python
## Python
pbp_py_run.down =\
    pbp_py_run.down.astype(str)

expected_yards_py =\
    smf.ols(
        data=pbp_py_run,
        formula="rushing_yards ~ 1 + down + ydstogo + " +
        "down:ydstogo + yardline_100 + " +
        "run_location + score_differential")\
        .fit()

pbp_py_run["ryoe"] =\
    expected_yards_py.resid
```

We include line breaks in our code to make it fit on the pages of this book. For example, in Python, we use the string "rush ing_yards ~ 1 + down + ydstogo + " followed by + and then a line break. Then, each of the next two strings, "down:ydstogo + yardline_100 + " and "run_location + score_differential" gets its own line breaks, and we then use the + character to *add* the strings together. These line breaks are not required but help make the code look better to human eyes and ideally be more readable.

Likewise, fit the model in, and save the residuals as RYOE in R:

```
## R
pbp_r_run <-
    pbp_r_run |>
    mutate(down = as.character(down))

expected_yards_r <-
    lm(rushing_yards ~ 1 + down + ydstogo + down:ydstogo +
        yardline_100 + run_location + score_differential,
        data = pbp_r_run
    )

pbp_r_run <-
    pbp_r_run |>
    mutate(ryoe = resid(expected_yards_r))
```

Now, examine the summary of the model in Python:

```
## Python
print(expected_yards_py.summary())
```

Resulting in:

```
                        OLS Regression Results
===============================================================================
Dep. Variable:          rushing_yards   R-squared:                      0.016
Model:                            OLS   Adj. R-squared:                 0.016
Method:                 Least Squares   F-statistic:                    136.6
Date:                Sun, 04 Jun 2023   Prob (F-statistic):          3.43e-313
Time:                        09:36:43   Log-Likelihood:              -2.9764e+05
No. Observations:               91442   AIC:                          5.953e+05
Df Residuals:                   91430   BIC:                          5.954e+05
Df Model:                          11
Covariance Type:            nonrobust
===============================================================================
                       coef     std err        t      P>|t|    [0.025    0.975]
-------------------------------------------------------------------------------
Intercept            1.6085       0.136   11.849      0.000     1.342     1.875
down[T.2.0]          1.6153       0.153   10.577      0.000     1.316     1.915
down[T.3.0]          1.2846       0.161    7.990      0.000     0.969     1.600
down[T.4.0]          0.2844       0.249    1.142      0.254    -0.204     0.773
run_location[T.middle]
```

	-0.5634	0.053	-10.718	0.000	-0.666	-0.460
run_location[T.right]						
	-0.0382	0.049	-0.784	0.433	-0.134	0.057
ydstogo	0.2024	0.014	14.439	0.000	0.175	0.230
down[T.2.0]:ydstogo						
	-0.1466	0.016	-8.957	0.000	-0.179	-0.115
down[T.3.0]:ydstogo						
	-0.0437	0.019	-2.323	0.020	-0.081	-0.007
down[T.4.0]:ydstogo						
	0.2302	0.090	2.567	0.010	0.054	0.406
yardline_100	0.0186	0.001	21.230	0.000	0.017	0.020
score_differential						
	-0.0040	0.002	-2.023	0.043	-0.008	-0.000

```
==============================================================================
Omnibus:                    80510.527   Durbin-Watson:                  1.979
Prob(Omnibus):                  0.000   Jarque-Bera (JB):         3941200.520
Skew:                           4.082   Prob(JB):                        0.00
Kurtosis:                      34.109   Cond. No.                        838.
==============================================================================
```

Notes:
[1] Standard Errors assume that the covariance matrix of the errors
is correctly specified.

Or examine the summary of the model in R:

```
## R
print(summary(expected_yards_r))
```

Resulting in:

```
Call:
lm(formula = rushing_yards ~ 1 + down + ydstogo + down:ydstogo +
    yardline_100 + run_location + score_differential, data = pbp_r_run)

Residuals:
    Min      1Q  Median      3Q     Max
-32.233  -3.130  -1.173   1.410  94.112

Coefficients:
                    Estimate Std. Error t value Pr(>|t|)
(Intercept)         1.608471   0.135753  11.849  < 2e-16 ***
down2               1.615277   0.152721  10.577  < 2e-16 ***
down3               1.284560   0.160775   7.990 1.37e-15 ***
down4               0.284433   0.249106   1.142   0.2535
ydstogo             0.202377   0.014016  14.439  < 2e-16 ***
yardline_100        0.018576   0.000875  21.230  < 2e-16 ***
run_locationmiddle -0.563369   0.052565 -10.718  < 2e-16 ***
run_locationright  -0.038176   0.048684  -0.784   0.4329
score_differential -0.004028   0.001991  -2.023   0.0431 *
down2:ydstogo      -0.146602   0.016367  -8.957  < 2e-16 ***
down3:ydstogo      -0.043703   0.018814  -2.323   0.0202 *
down4:ydstogo       0.230179   0.089682   2.567   0.0103 *
---
```

```
Signif. codes:  0 '***' 0.001 '**' 0.01 '*' 0.05 '.' 0.1 ' ' 1

Residual standard error: 6.272 on 91430 degrees of freedom
Multiple R-squared:  0.01617,   Adjusted R-squared:  0.01605
F-statistic: 136.6 on 11 and 91430 DF,  p-value: < 2.2e-16
```

Each estimated coefficient helps you tell a story about rushing yards during plays:

- Running plays on the second down (down[T.2.0] in Python or down2 in R) have more expected yards per carry than the first down, all else being equal—in this case, about 1.6 yards.

- Running plays on the third down (down[T.3.0] in Python or down3 in R) have more expected yards per carry than the first down, all else being equal—in this case, about 1.3 yards.

- The interaction term tells you that this is especially true when fewer yards to go remain for the first down (the interaction terms are all negative). From a football standpoint, this just means that second and third downs, and short yardage to go for a first down or touchdown, are more favorable to the offense running the ball than first down and 10 yards to go.

- Conversely, running plays on fourth down have slightly more yards gained compared to first down (down[T.4.0] in Python or down4 in R), all else being equal, but not as many as second or third down.

- As the number of yards to go increases (ydstogo), so do the rushing yards, all else being equal, with each yard to go worth about a fifth of a yard. This is because the ydstogo estimate is positive.

- As the ball is farther from the endzone, rushing plays produce slightly more yards per play (about 0.02 per yard to go to the endzone; yardline_100). For example, even if the team has 100 yards to go, a coefficient of 0.02 means that only 2 extra yards would be rushed in the play, which doesn't have a big impact compared to other coefficients.

- Rushing plays in the middle of the field earn about a half yard less than plays to the left, based on the contrast estimate between the middle and left side of the field (run_location[T.middle] in Python or run_locationmiddle in R).

- The negative score_differential coefficient differs statistically from 0. Thus, when teams are ahead (have a positive score differential), they gain fewer yards per play on the average running play. However, this effect is so small (0.004) and thus not important compared to other coefficients that it can be ignored (for example, being up by 50 points would decrease the number of yards by only 0.2 per carry).

Notice from the coefficients that running to the middle of the field is harder than running to the outside, all other factors being equal. You indeed see that distance and yards to go to the first down marker and to the endzone both positively affect rushing yards: the farther an offensive player is away from the goal, the more defense will surrender to that offensive player on average.

The kableExtra package in R helps produce well-formatted tables in R Markdown and Quarto documents as well as onscreen. You'll need to install the package if you have not done so already.

Tables provide another way to present regression coefficients. For example, the broom package allows you to create tidy tables in R that can then be formatted by using the kableExtra package, such as Table 4-1. Specifically, with this code, take the model fit expected_yards_r and then pipe the model to extract the *tidy* model outputs (including the 95% CIs) by using tidy(conf.int = TRUE). Then, convert the table to a kable table and show two digits by using kbl(format = "pipe", digits = 2). Last, apply styling from the kableExtra package by using kable_styling():

```
## R
library(broom)
library(kableExtra)
expected_yards_r |>
    tidy(conf.int = TRUE) |>
    kbl(format = "pipe", digits = 2) |>
    kable_styling()
```

Table 4-1. Example regression coefficient table. term is the regression coefficient, estimate is the estimated value for the coefficient, std.error is the standard error, statistic is the t-score, p.value is the p-value, conf.low is the bottom of the 95% CI, and conf.high is the top of the 95% CI.

Term	estimate	std.error	statistic	p.value	conf.low	conf.high
(Intercept)	1.61	0.14	11.85	0.00	1.34	1.87
down2	1.62	0.15	10.58	0.00	1.32	1.91
down3	1.28	0.16	7.99	0.00	0.97	1.60
down4	0.28	0.25	1.14	0.25	-0.20	0.77
ydstogo	0.20	0.01	14.44	0.00	0.17	0.23
yardline_100	0.02	0.00	21.23	0.00	0.02	0.02
run_locationmiddle	-0.56	0.05	-10.72	0.00	-0.67	-0.46
run_locationright	-0.04	0.05	-0.78	0.43	-0.13	0.06
score_differential	0.00	0.00	-2.02	0.04	-0.01	0.00
down2:ydstogo	-0.15	0.02	-8.96	0.00	-0.18	-0.11

Term	estimate	std.error	statistic	p.value	conf.low	conf.high
down3:ydstogo	-0.04	0.02	-2.32	0.02	-0.08	-0.01
down4:ydstogo	0.23	0.09	2.57	0.01	0.05	0.41

 Writing about regressions can be hard, and knowing your audience and their background is key. For example, the paragraph "Notice from the coefficients…" would be appropriate for a casual blog but not for a peer-reviewed sports journal. Likewise, a table like Table 4-1 might be included in a journal article but is more likely to be included as part of a technical report or journal article's supplemental materials. Our walk-through of individual coefficients in a bulleted list might be appropriate for a report to a client who wants an item-by-item description or for a teaching resource such as a blog or book on football analytics.

Analyzing RYOE

Just as with the first version of RYOE in Chapter 3, now you will analyze the new version of your metric for rushers. First, create the summary tables for the RYOE totals, means, and yards per carry in Python. Next, save only data from players with more than 50 carries. Also, rename the columns and sort by total RYOE:

```
## Python
ryoe_py =\
    pbp_py_run\
    .groupby(["season", "rusher_id", "rusher"])\
    .agg({
        "ryoe": ["count", "sum", "mean"],
        "rushing_yards": ["mean"]})

ryoe_py.columns =\
    list(map("_".join, ryoe_py.columns))
ryoe_py.reset_index(inplace=True)

ryoe_py =\
    ryoe_py\
    .rename(columns={
        "ryoe_count": "n",
        "ryoe_sum": "ryoe_total",
        "ryoe_mean": "ryoe_per",
        "rushing_yards_mean": "yards_per_carry"
    })\
    .query("n > 50")

print(ryoe_py\
    .sort_values("ryoe_total", ascending=False)
    )
```

Resulting in:

```
        season  rusher_id        rusher  ...  ryoe_total  ryoe_per  yards_per_carry
1870      2021  00-0036223     J.Taylor  ...  471.232840  1.419376         5.454819
1350      2020  00-0032764      D.Henry  ...  345.948778  0.875820         5.232911
1183      2019  00-0034796   L.Jackson  ...  328.524757  2.607339         6.880952
1069      2019  00-0032764      D.Henry  ...  311.641243  0.807361         5.145078
1383      2020  00-0033293      A.Jones  ...  301.778866  1.365515         5.565611
 ...       ...        ...          ...  ...         ...       ...              ...
627       2018  00-0027029      L.McCoy  ... -208.392834 -1.294365         3.192547
51        2016  00-0027155   R.Jennings  ... -228.084591 -1.226261         3.344086
629       2018  00-0027325     L.Blount  ... -235.865233 -1.531592         2.714286
991       2019  00-0030496       L.Bell  ... -338.432836 -1.381359         3.220408
246       2016  00-0032241     T.Gurley  ... -344.314622 -1.238542         3.183453

[533 rows x 7 columns]
```

Next, sort by mean RYOE per carry in Python:

```
## Python
print(
    ryoe_py\
    .sort_values("ryoe_per", ascending=False)
    )
```

Resulting in:

```
        season  rusher_id          rusher  ...  ryoe_total  ryoe_per  yards_per_carry
2103      2022  00-0034796     L.Jackson  ...  280.752317  3.899338         7.930556
1183      2019  00-0034796     L.Jackson  ...  328.524757  2.607339         6.880952
1210      2019  00-0035228     K.Murray   ...  137.636412  2.596913         6.867925
2239      2022  00-0036945     J.Fields   ...  177.409631  2.304021         6.506494
1467      2020  00-0034796     L.Jackson  ...  258.059489  2.186945         6.415254
 ...       ...        ...            ...  ...         ...       ...              ...
1901      2021  00-0036414       C.Akers  ... -129.834294 -1.803254         2.430556
533       2017  00-0032940  D.Washington  ... -105.377929 -1.848736         2.684211
1858      2021  00-0035860       T.Jones  ... -100.987077 -1.870131         2.629630
60        2016  00-0027791      J.Starks  ... -129.298259 -2.052353         2.301587
1184      2019  00-0034799     K.Ballage  ... -191.983153 -2.594367         1.824324

[533 rows x 7 columns]
```

These same tables may be created in R as well:

```
## R
ryoe_r <-
    pbp_r_run |>
    group_by(season, rusher_id, rusher) |>
    summarize(
        n = n(), ryoe_total = sum(ryoe), ryoe_per = mean(ryoe),
        yards_per_carry = mean(rushing_yards)
    ) |>
    filter(n > 50)
```

```
ryoe_r |>
    arrange(-ryoe_total) |>
    print()
```

Resulting in:

```
# A tibble: 533 × 7
# Groups:   season, rusher_id [533]
   season rusher_id  rusher         n ryoe_total ryoe_per yards_per_carry
    <dbl> <chr>      <chr>      <int>      <dbl>    <dbl>           <dbl>
 1   2021 00-0036223 J.Taylor     332       471.    1.42            5.45
 2   2020 00-0032764 D.Henry      395       346.    0.876           5.23
 3   2019 00-0034796 L.Jackson    126       329.    2.61            6.88
 4   2019 00-0032764 D.Henry      386       312.    0.807           5.15
 5   2020 00-0033293 A.Jones      221       302.    1.37            5.57
 6   2022 00-0034796 L.Jackson     72       281.    3.90            7.93
 7   2019 00-0031687 R.Mostert    190       274.    1.44            5.83
 8   2016 00-0033045 E.Elliott    342       274.    0.800           5.14
 9   2020 00-0034796 L.Jackson    118       258.    2.19            6.42
10   2021 00-0034791 N.Chubb      228       248.    1.09            5.52
# i 523 more rows
```

Then sort by mean RYOE per carry in R:

```
## R
ryoe_r |>
    filter(n > 50) |>
    arrange(-ryoe_per) |>
    print()
```

Resulting in:

```
# A tibble: 533 × 7
# Groups:   season, rusher_id [533]
   season rusher_id  rusher         n ryoe_total ryoe_per yards_per_carry
    <dbl> <chr>      <chr>      <int>      <dbl>    <dbl>           <dbl>
 1   2022 00-0034796 L.Jackson     72       281.    3.90            7.93
 2   2019 00-0034796 L.Jackson    126       329.    2.61            6.88
 3   2019 00-0035228 K.Murray      53       138.    2.60            6.87
 4   2022 00-0036945 J.Fields      77       177.    2.30            6.51
 5   2020 00-0034796 L.Jackson    118       258.    2.19            6.42
 6   2017 00-0027939 C.Newton      92       191.    2.08            6.17
 7   2020 00-0035228 K.Murray      70       144.    2.06            6.06
 8   2021 00-0034750 R.Penny      119       242.    2.03            6.29
 9   2019 00-0034400 J.Wilkins     51       97.8    1.92            6.02
10   2022 00-0033357 T.Hill        95       171.    1.80            6.05
# i 523 more rows
```

The preceding outputs are similar to those from Chapter 3 but come from a model that *corrects for* additional features when estimating RYOE.

When it comes to total RYOE, Jonathan Taylor's 2021 season still reigns supreme, and now by over 100 yards over expected more than the next-best player, while Derrick Henry makes a couple of appearances again. Nick Chubb has been one of the best runners in the entire league since he was drafted in the second round in 2018, while Raheem Mostert of the 2019 NFC champion San Francisco 49ers wasn't even drafted at all but makes the list. Ezekiel Elliott of the Cowboys and Le'Veon Bell of the Pittsburgh Steelers, New York Jets, Chiefs, and Ravens make appearances at various places: we'll talk about them later.

As for RYOE per carry (for players with 50 or more carries in a season), Rashaad Penny's brilliant 2021 shines again. But the majority of the players in this list are quarterbacks, like 2019 league MVP Lamar Jackson; 2011 and 2019 first-overall picks Cam Newton and Kyler Murray, respectively; and the New Orleans Saints' Taysom Hill. Justin Fields, who had one of the best rushing seasons in the history of the quarterback position in 2022, shows up as well.

As for the stability of this metric relative to yards per carry, recall that in the previous chapter, you got a slight bump in stability though not necessarily enough to say that RYOE per carry is definitely a superior metric to yards per carry for rushers. Let's redo this analysis:

```python
## Python
# keep only the columns needed
cols_keep =\
    ["season", "rusher_id", "rusher",
     "ryoe_per", "yards_per_carry"]

# create current dataframe
ryoe_now_py =\
    ryoe_py[cols_keep].copy()

# create last-year's dataframe
ryoe_last_py =\
    ryoe_py[cols_keep].copy()

# rename columns
ryoe_last_py\
    .rename(columns = {'ryoe_per': 'ryoe_per_last',
                       'yards_per_carry': 'yards_per_carry_last'},
            inplace=True)

# add 1 to season
ryoe_last_py["season"] += 1

# merge together
ryoe_lag_py =\
    ryoe_now_py\
    .merge(ryoe_last_py,
           how='inner',
           on=['rusher_id', 'rusher',
               'season'])
```

Then examine the correlation for yards per carry:

```Python
## Python
ryoe_lag_py[["yards_per_carry_last", "yards_per_carry"]]\
    .corr()
```

Resulting in:

```
                     yards_per_carry_last  yards_per_carry
yards_per_carry_last             1.000000         0.347267
yards_per_carry                  0.347267         1.000000
```

Repeat with RYOE:

```Python
## Python
ryoe_lag_py[["ryoe_per_last", "ryoe_per"]]\
    .corr()
```

Resulting in:

```
               ryoe_per_last  ryoe_per
ryoe_per_last       1.000000  0.373582
ryoe_per            0.373582  1.000000
```

These calculations may also be done using R:

```R
## R
# create current dataframe
ryoe_now_r <-
    ryoe_r |>
    select(-n, -ryoe_total)

# create last-year's dataframe
# and add 1 to season
ryoe_last_r <-
    ryoe_r |>
    select(-n, -ryoe_total) |>
    mutate(season = season + 1) |>
    rename(ryoe_per_last = ryoe_per,
           yards_per_carry_last = yards_per_carry)

# merge together
ryoe_lag_r <-
    ryoe_now_r |>
    inner_join(ryoe_last_r,
               by = c("rusher_id", "rusher", "season")) |>
    ungroup()
```

Then select the two yards-per-carry columns and examine the correlation:

```R
## R
ryoe_lag_r |>
    select(yards_per_carry, yards_per_carry_last) |>
    cor(use = "complete.obs")
```

Resulting in:

```
                     yards_per_carry yards_per_carry_last
yards_per_carry             1.000000             0.347267
yards_per_carry_last        0.347267             1.000000
```

Repeat with the RYOE columns:

```R
## R
ryoe_lag_r |>
    select(ryoe_per, ryoe_per_last) |>
    cor(use = "complete.obs")
```

Resulting in:

```
                ryoe_per ryoe_per_last
ryoe_per       1.0000000     0.3735821
ryoe_per_last  0.3735821     1.0000000
```

This is an interesting result! The year-to-year stability for RYOE slightly improves with the new model, meaning that after stripping out more and more context from the situation, we are indeed extracting some additional signal in running back ability above expectation. The issue is that this is still a minimal improvement, with the difference in r values less than 0.03. Additional work should interrogate the problem further, using better data (like tracking data) and better models (like tree-based models).

 You have hopefully noticed that the code in this chapter and in Chapter 3 are repetitive and similar. If we were repeating code on a regular basis like this, we would write our own set of functions and place them in a package to allow us to easily reuse our code. "Packages" on page 255 provides an overview of this topic.

So, Do Running Backs Matter?

This question seems silly on its face. Of course running backs, the players who carry the ball 50 to 250 times a year, matter. Jim Brown, Franco Harris, Barry Sanders, Marshall Faulk, Adrian Peterson, Derrick Henry, Jim Taylor—the history of the NFL cannot be written without mentioning the feats of great runners.

Ben Baldwin hosts a Nerd-to-Human Translator (*https://oreil.ly/ L2YJ6*), which helps describe our word choices in this chapter. He notes, "What the nerds say: running backs don't matter." And then, "What the nerds mean: the results of running plays are primarily determined by run blocking and defenders in the box, not who is carrying the ball. Running backs are interchangeable, and investing a lot of resources (in the draft or free agency) doesn't make sense." And then he provides supporting evidence. Check out his page for this evidence and more useful tips.

However, we should note a couple of things. First, passing the football has gotten increasingly easier over the course of the past few decades. In "The Most Important Job in Sports Is Easier Than Ever Before" (*https://oreil.ly/RVXv3*), Kevin Clark notes how "scheme changes, rule changes, and athlete changes" all help quarterbacks pass for more yards. We'll add technology (e.g., the gloves the players wear), along with the adoption of college offenses and the creative ways in which they pass the football, to Kevin's list.

In other words, the notion that "when you pass, only three things can happen (a completion, an incompletion, or an interception), and two of them are bad" doesn't take into account the increasing rate at which the first thing (a completion) happens. Passing has long been more efficient than running (see "Exercises" on page 111), but now the variance in passing the ball has shrunk enough to prefer it in most cases to the low variance (and low efficiency) of running the ball.

In addition to the fact that running the ball is less efficient than passing the ball, whether measured through yards per play or something like expected points added (EPA), the player running the ball evidently has less influence over the outcome of a running play than previously thought.

Other factors that we couldn't account for by using `nflfastR` data also work against running backs influencing their production. Eric, while he was at PFF (*https://oreil.ly/ 4i_pH*), showed that offensive line play carried a huge signal when it came to rushing plays, using PFF's player grades by offensive linemen on a given play.

Eric also showed during the 2021 season (*https://oreil.ly/tqFn7*) that the concept of a perfectly blocked run—running play where no run blocker made a negative play (such as getting beaten by a defender)—changed the outcome of a running play by roughly *half of an expected point*. While the running back can surely aid in the creation of a perfectly blocked play through setting up his blockers and the like, the magnitude of an individual running back's influence is unlikely to be anywhere close to this value.

The NFL has generally caught up to this phenomenon, spending less and less as a percentage of the salary cap on running backs for the past decade plus, as Benjamin Morris shows in "Running Backs Are Finally Getting Paid What They're Worth" (*https://oreil.ly/7j9Vw*) from FiveThirtyEight. Be that as it may, some contention remains about the position, with some players who are said to "break the mold" getting big contracts, usually to the disappointment of their teams.

An example of a team overpaying its star running back, and likely regretting it, occurred when the Dallas Cowboys in the fall of 2019, signed Ezekiel Elliott to a six-year, $90 million contract, with $50 million in guarantees. Elliott was holding out for a new deal, no doubt giving Cowboys owner Jerry Jones flashbacks to 1993, when the NFL's all-time leading rusher Emmitt Smith held out for the first two games for the defending champion Cowboys. The Cowboys, after losing both games without Smith, caved to the star's demands, and he rewarded them by earning the NFL's MVP award during the regular season. He finished off the season by earning the Super Bowl MVP as well, helping lead Dallas to the second of its three championships in the mid-'90s.

Elliott had had a similar start to his career as Smith, leading the NFL in rushing yards per game in each of his first three seasons leading up to his holdout, and leading the league in total rushing yards during both seasons when he played the full year (Elliott was suspended for parts of the 2017 season). The Cowboys won their division in 2016 and 2018, and won a playoff game, only its second since 1996, in 2018.

As predicted by many in the analytics community, Zeke struggled to live up to the deal, with his rushing yards per game falling from 84.8 in 2019, to 65.3 in 2020, and flattening out to 58.9 and 58.4 in 2021 and 2022, respectively. Elliott's yards per carry in 2022 fell beneath 4.0 to 3.8, and he eventually lost his starting job—and job altogether—to his backup Tony Pollard, in the 2023 offseason.

Not only did Elliott fail to live up to his deal, but his contract was onerous enough that the Cowboys had to jettison productive players like wide receiver Amari Cooper to get under the salary cap, a cascading effect that weakened the team more than simply the negative surplus play of a running back could.

An example of a team holding its ground, and likely breathing a sigh of relief, occurred in 2018. Le'Veon Bell, who showed up in "Analyzing RYOE" on page 100, held out of training camp and the regular season for the Pittsburgh Steelers in a contract dispute, refusing to play on the *franchise tag*—an instrument a team uses to keep a player around for one season at the average of the top five salaries at their respective position. Players, who usually want longer-term deals, will often balk at playing under this contract type, and Bell—who gained over 5,000 yards on the ground during his first four years with the club, and led the NFL in touches in 2017 (carries and receptions)—was one such player.

The problem for Bell was that he was easily replaced in Pittsburgh. James Conner, a cancer survivor from the University of Pittsburgh selected in the third round of the 2017 NFL Draft, rushed for over 950 yards, at 4.5 yards per carry (Bell's career average was 4.3 as a Steeler), with 13 total touchdowns—earning a Pro Bowl berth.

Bell was allowed to leave for the lowly Jets the following year, where he lasted 17 games and averaged just 3.3 yards per carry, scoring just four times. He did land with the Kansas City Chiefs after being released during the 2020 season. While that team made the Super Bowl, he didn't play in that game, and the Chiefs lost 31–9 to Tampa Bay. He was out of football after the conclusion of the 2021 season.

Elliott's and Bell's stories are maybe the most dramatic falls from grace for a starting running back but are not the only ones, which is why it's important to be able to correctly attribute production to both the player and the scheme/rest of the roster in proper proportions, so as not to make a poor investment salary-wise.

Assumption of Linearity

Figures 4-7 and 4-8 clearly show a nonlinear relationship. Technically, linear regression assumes that both the residuals are normally distributed and that the observed relationship is linear. However, the two usually go hand in hand.

Base R contains useful diagnostic tools for looking at the results of linear models. The tools use base R's `plot()` function (this is one of the few times we like `plot()`; it creates a simple function that we use only for internal diagnosis, not to share with others). First, use `par(mfrow=c(2,2))` to create four subplots. Then `plot()` the multiple regression you previously fit and saved as `expected_yards_r` to create Figure 4-13:

```
## R
par(mfrow = c(2, 2))
plot(expected_yards_r)
```

Figure 4-13 contains four subplots. The top left shows the model's estimated (or fitted) values compared to the difference in predicted versus fitted (or residual). The top right shows cumulative distribution of the model's fits against the theoretical values. The bottom left shows the fitted values versus the square root of the standardized residuals. The bottom right shows the influence of a parameter on the model's fit against the standardized residual.

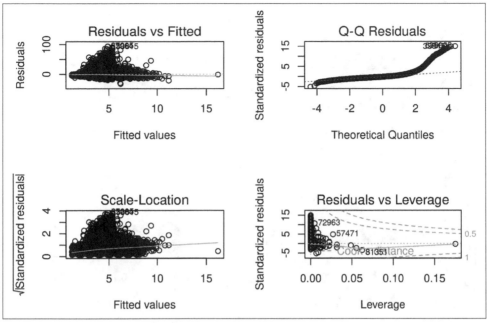

Figure 4-13. Four diagnostic subplots for a regression model

The "Residuals vs. Fitted" subplot does not look *too* pathological. This subplot simply shows that many data points do not fit the model well and that a skew exists with the data. The "Normal Q-Q" subplot shows that many data points start to diverge from the expected model. Thus, our model does not fit the data well in some cases. The "Scale-Location" subplot shows similar patterns as the "Residuals vs. Fitted." Also, this plot has a weird pattern with *W*-like lines near 0 due to the integer (e.g., 0, 1, 2, 3) nature of the data. Lastly, "Residuals vs. Leverage" shows that some data observations have a great deal of "leverage" on the model's estimates, but these fall within the expected range based on their Cook's distance. *Cook's distance* is, informally, the estimated influence of an observation on the model's fit. Basically, a larger value implies that an observation has a greater effect on the model's estimates.

Let's look at what happens if you remove plays less than 15 yards or greater than 90 to create Figure 4-14:

```R
## R
expected_yards_filter_r <-
    pbp_r_run |>
    filter(rushing_yards > 15 & rushing_yards < 90) |>
    lm(formula = rushing_yards ~ 1 + down + ydstogo + down:ydstogo +
                yardline_100 + run_location + score_differential)

par(mfrow = c(2, 2))
plot(expected_yards_filter_r)
```

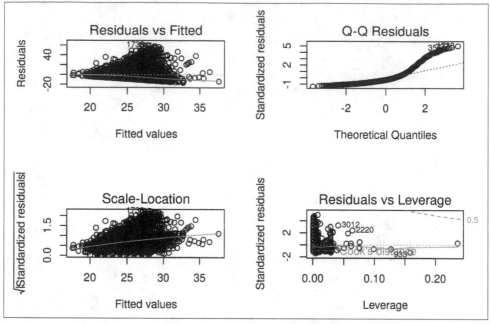

Figure 4-14. This figure is a re-creation of Figure 4-13 with the model including only rushing plays with more than 15 yards and less than 95 yards

Figure 4-14 shows that the new linear model, `expected_yards_filter_r`, fits better. Although the "Residuals vs. Fitted" subplot has a wonky straight line (reflecting that the data has now been censored), the other subplots look better. The most improved subplot is "Normal Q-Q." Figure 4-13 has a scale from –5 to 15, whereas this plot now has a scale from –1 to 5.

As one last check, look at the summary of the model and notice the improved model fit. The R^2 value improved from ~0.01 to 0.05:

```
## R
summary(expected_yards_filter_r)
```

Which results in:

```
Call:
lm(formula = rushing_yards ~ 1 + down + ydstogo + down:ydstogo +
    yardline_100 + run_location + score_differential, data = filter(pbp_r_run,
    rushing_yards > 15 & rushing_yards < 95))


Residuals:
    Min      1Q  Median      3Q     Max
-17.158  -7.795  -3.766   3.111  63.471
```

```
Coefficients:
                     Estimate Std. Error t value Pr(>|t|)
(Intercept)          21.950963   2.157834  10.173   <2e-16 ***
down2                -2.853904   2.214676  -1.289   0.1976
down3                -0.696781   2.248905  -0.310   0.7567
down4                 0.418564   3.195993   0.131   0.8958
ydstogo              -0.420525   0.204504  -2.056   0.0398 *
yardline_100          0.130255   0.009975  13.058   <2e-16 ***
run_locationmiddle    0.680770   0.562407   1.210   0.2262
run_locationright     0.635015   0.443208   1.433   0.1520
score_differential    0.048017   0.019098   2.514   0.0120 *
down2:ydstogo         0.207071   0.224956   0.920   0.3574
down3:ydstogo         0.165576   0.234271   0.707   0.4798
down4:ydstogo         0.860361   0.602634   1.428   0.1535
---
Signif. codes:  0 '***' 0.001 '**' 0.01 '*' 0.05 '.' 0.1 ' ' 1

Residual standard error: 12.32 on 3781 degrees of freedom
Multiple R-squared:  0.05074,   Adjusted R-squared:  0.04798
F-statistic: 18.37 on 11 and 3781 DF,  p-value: < 2.2e-16
```

In summary, looking at the model's residuals helped you to see that the model does not do well for plays shorter than 15 yards or longer than 95 yards. Knowing and quantifying this limitation at least helps you to know what you do not know with your model.

Data Science Tools Used in This Chapter

This chapter covered the following topics:

- Fitting a multiple regression in Python by using OLS() or in R by using lm()
- Understanding and reading the coefficients from a multiple regression
- Reapplying data-wrangling tools you learned in previous chapters
- Examining a regression's fit

Exercises

1. Change the carries threshold from 50 carries to 100 carries. Do you still see the stability differences that you found in this chapter?

2. Use the full nflfastR dataset to show that rushing is less efficient than passing, both using yards per play and EPA per play. Also inspect the variability of these two play types.

3. Is rushing more valuable than passing in some situations (e.g., near the opposing team's end zone)?

4. Inspect James Conner's RYOE values for his career relative to Bell's. What do you notice about the metric for both backs?

5. Repeat the processes within this chapter with receivers in the passing game. To do this, you have to filter by `play_type == "pass"` and `receiver_id` not being `NA` or `NULL`. Finding features will be difficult, but consider the process in this chapter for guidance. For example, use `down` and `distance`, but maybe also use something like `air_yards` in your model to try to set an expectation.

Suggested Readings

The books listed in "Suggested Readings" on page 77 also apply for this chapter. Building upon this list, here are some other resources you may find helpful:

- *The Chicago Guide to Writing about Numbers*, 2nd edition, by Jane E. Miller (Chicago Press, 2015) provides great examples for describing numbers in different forms of writing.

- *The Chicago Guide to Writing about Multivariate Analysis*, 2nd edition, by Jane E. Miller (University of Chicago Press, 2013) provides many examples describing multiple regression. Although we disagree with her use of *multivariate regression* as a synonym for *multiple regression*, the book does a great job proving examples of describing regression outputs.

- *Practical Linear Algebra for Data Science* by Mike X Cohen (O'Reilly Media, 2022) provides understanding of linear algebra that will help you better understand regression. Linear algebra forms the foundation of almost all statistical methods including multiple regression.

- FiveThirtyEight (*https://fivethirtyeight.com*) contains a great deal of data journalism and was started and run by Nate Silver, until he exited from the ABC/Disney-owned site in 2023. Look through the posts and try to tell where the site uses regression models for posts.

- *Statistical Modeling, Causal Inference, and Social Science* (*https://statmodeling.stat.columbia.edu*), created by Andrew Gelman with contributions from many other authors, is a blog that often discusses regression modeling. Gelman is a more academic political scientist version of Nate Silver.

- *Statistical Thinking* (*https://www.fharrell.com*) by Frank Harrell is a blog that also commonly discusses regression analysis. Harrell is a more academic statistician version of Nate Silver who usually focuses more on statistics. However, many of his posts are often relevant to people doing any type of regression analysis.

## CHAPTER 5

Generalized Linear Models: Completion Percentage over Expected

In Chapters 3 and 4, you used both simple and multiple regression to adjust play-by-play data for the *context* of the play. In the case of ball carriers, you adjusted for the situation (such as down, distance, yards to go) to calibrate individual player statistics on the play level, and later the season level. This approach clearly can be applied to the passing game, and more specifically, quarterbacks. As discussed in Chapter 3, Minnesota quarterback Sam Bradford set the NFL record for seasonal completion percentage in 2016, completing a whopping 71.6% of his passes.

Bradford, however, was just a middle-of-the-pack quarterback in terms of efficiency—whether measured by yards per pass attempt, expected points per passing attempt, or touchdown passes. The Vikings won only 7 of his 15 starts that year. The reason Bradford's completion percentage was so high was that he averaged just 6.6 yards for depth per target (37th in the NFL, per PFF). In general, passes that are thrown longer distances are completed at a lower rate.

To see this, you will create Figure 5-1 in Python or Figure 5-2 in R. First, load the data. Then, filter pass plays (`play_type == "pass"`) with a passer (`passer_id.not null()` in Python or `!is.na(passer_id)` in R), and a pass depth (`air_yards.not null()` in Python or `!is.na(air_yards)` in R). In Python, use this code:

```python
## Python
import pandas as pd
import numpy as np
import nfl_data_py as nfl
import statsmodels.formula.api as smf
import statsmodels.api as sm
import matplotlib.pyplot as plt
import seaborn as sns
```

```
seasons = range(2016, 2022 + 1)
pbp_py = nfl.import_pbp_data(seasons)

pbp_py_pass = \
    pbp_py\
    .query('play_type == "pass" & passer_id.notnull() &' +
           'air_yards.notnull()')\
    .reset_index()
```

Or with R, use this code:

```
## R
library(tidyverse)
library(nflfastR)
library(broom)

pbp_r <- load_pbp(2016:2022)
pbp_r_pass <-
    pbp_r |>
    filter(play_type == "pass" & !is.na(passer_id) &
           !is.na(air_yards))
```

Next, restrict air yards to be greater than 0 yards and less than or equal to 20 yards in order to ensure that you have a large enough sample size. Summarize the data to calculate the completion percentage, comp_pct. Then plot results to create Figure 5-1:

```
## Python
# Change theme for chapter
sns.set_theme(style="whitegrid", palette="colorblind")

# Format and then plot
pass_pct_py = \
    pbp_py_pass\
    .query('0 < air_yards <= 20')\
    .groupby('air_yards')\
    .agg({"complete_pass": ["mean"]})

pass_pct_py.columns = \
    list(map('_'.join, pass_pct_py.columns))

pass_pct_py\
    .reset_index(inplace=True)
pass_pct_py\
    .rename(columns={'complete_pass_mean': 'comp_pct'},
            inplace=True)

sns.regplot(data=pass_pct_py, x='air_yards', y='comp_pct',
            line_kws={'color': 'red'});
plt.show();
```

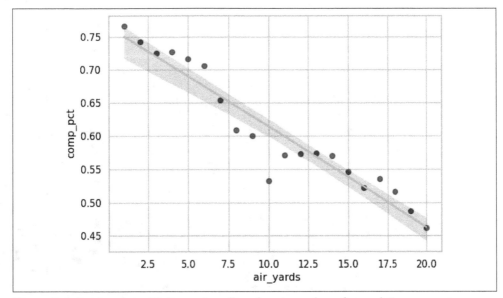

Figure 5-1. Scatterplot with linear trendline for air yards and completion percentage, plotted with seaborn

Or use R to create Figure 5-2:

```
## R
pass_pct_r <-
  pbp_r_pass |>
  filter(0 < air_yards & air_yards <= 20) |>
  group_by(air_yards) |>
  summarize(comp_pct = mean(complete_pass),
            .groups = 'drop')

pass_pct_r |>
  ggplot(aes(x = air_yards, y=comp_pct)) +
  geom_point() +
  stat_smooth(method='lm') +
  theme_bw() +
  ylab("Percent completion") +
  xlab("Air yards")
```

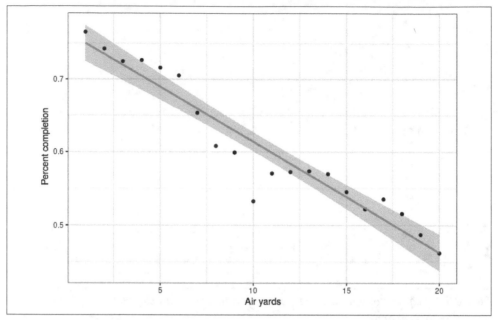

Figure 5-2. Scatterplot with linear trendline for air yards and completion percentage, plotted with ggplot2

Figures 5-1 and 5-2 clearly show a trend, as expected. Thus, any discussion of quarterback accuracy—as measured by completion percentage—needs to be accompanied by some adjustment for style of play.

Completion percentage over expected, referred to as *CPOE* in the football analytics world (and introduced in Chapter 1), is one of the adjusted metrics that has made its way into the mainstream. Ben Baldwin's website (*https://rbsdm.com*), a great reference for open football data, displays CPOE as one of its main metrics in large part because CPOE together with EPA per passing play has shown to be the most predictive public metric for quarterback play from year to year. The NFL's Next Generation Stats (NGS) group (*https://nextgenstats.nfl.com*) has its own version of CPOE, which includes tracking data-engineered features like receiver separation from nearest defender, prominently displayed on its website. ESPN uses the metric consistently in its broadcasts.

Measuring quarterback performance this way has some issues, which we will touch on at the end of the chapter, but CPOE is here to stay. We will start the process of walking you through its development by using generalized linear models.

Generalized Linear Models

Chapter 3 defined and described some key assumptions of simple linear regression. These assumptions included the following:

- The predictor is linearly related to a single dependent variable, or feature.
- One predictor variable (simple linear regression) or more predictor variables (multiple regression) describe the dependent variable.

Another key assumption of multiple regression is that the distribution of the residuals follows a normal, or bell-curve, distribution. Although almost all datasets violate this last assumption, usually the assumption works "well enough." However, some data structures cause multiple regression to fail or produce nonsensical results. For example, completion percentage is a value bounded between 0 and 1 (*bounded* means the value cannot be smaller than 0 or larger than 1), as a pass is either incomplete (pass_complete = 0) or complete (pass_complete = 1). Hence a linear regression, which assumes no bounds on the response variable, is often inappropriate.

Likewise, other data commonly violates this assumption. For example, count data often has too many 0s to be normal and also cannot be negative (such as sacks per game). Likewise, binary data with two outcomes (such as win/lose or incomplete/complete pass) and discrete outcomes (such as passing location, which can be right, left, or middle) does not work with multiple regression as response data.

A class of regression models exists to model these types of outcomes: *generalized linear models* (*GLMs*). GLMs *generalize*, or extend, *linear models* to allow for response variables that are assumed to come from a non-normal distribution (such as binary responses or counts). The specific type of response distribution is called the *family*. One special type of GLM can be used to model binary data and is covered in this chapter. Chapter 6 covers how to use another type of GLM, the Poisson regression, with count data.

Other types of data can be analyzed by GLMs. For example, discrete outcomes can be analyzed using ordinal regression (also known as *ordinal classification*) but are beyond the scope of this book. Lastly, linear models (also known as *linear regression* and *ordinary least squares*) are a special type of a GLM—specifically, a GLM with a normal, or Gaussian, family.

To understand the basic theory behind how GLMs work, look at a completed pass that can be either 1 (completed) or 0 (incomplete). Because two outcomes are possible, this a *binary* response, and you can assume that a *binomial* distribution does a "good enough job" of describing the data. A normal distribution assumes two parameters: one for the center of the bell curve (the mean), and a second to describe the width of the bell curve (the standard deviation). In contrast, a binomial distribution requires only one parameter: a probability of success. With the pass example, this would be the probability of completing a pass. However, statistically modeling probability is hard because it is bounded by 0 and 1. So, a *link function* converts (or *links*) probability (a value ranging from 0 to 1) to a value ranging from $-\infty$ to ∞. The most common link function is the *logit*, which gives a name to one of the most common types of GLMs, the *logistic regression*.

Building a GLM

To apply GLMs, and specifically a logistic regression, you will start with a simple example. Let's begin by examining `air_yards` as our one feature for predicting completed passes. As suggested in Figures 5-1 and 5-2, longer passes are less likely to be completed. Now, you'll use a model to quantify this relation.

With Python, use the `glm()` function from `statsmodels.formula.api` (imported as `smf`) as well as the `binomial` family from `statsmodels.api` (imported as `sm`) with the play-by-play data to fit a GLM and then look at the model fit's summary:

```python
## Python
complete_ay_py = \
    smf.glm(formula='complete_pass ~ air_yards',
            data=pbp_py_pass,
            family=sm.families.Binomial())\
        .fit();

complete_ay_py.summary()
```

Resulting in:

```
<class 'statsmodels.iolib.summary.Summary'>
"""
                 Generalized Linear Model Regression Results
==============================================================================
Dep. Variable:          complete_pass   No. Observations:           131606
Model:                            GLM   Df Residuals:               131604
Model Family:                Binomial   Df Model:                        1
Link Function:                  Logit   Scale:                      1.0000
Method:                          IRLS   Log-Likelihood:            -81073.
Date:                Sun, 04 Jun 2023   Deviance:                1.6215e+05
Time:                        09:37:33   Pearson chi2:              1.32e+05
No. Iterations:                     5   Pseudo R-squ. (CS):        0.07013
Covariance Type:            nonrobust
==============================================================================
                 coef    std err          z      P>|z|      [0.025      0.975]
------------------------------------------------------------------------------
Intercept      1.0720      0.008    133.306      0.000       1.056       1.088
air_yards     -0.0573      0.001    -91.806      0.000      -0.059      -0.056
==============================================================================
"""
```

Likewise, with R, use the glm() function that is included with the core R packages, include a binomial family, and then look at the summary:

```
## R
complete_ay_r <-
  glm(complete_pass ~ air_yards,
      data = pbp_r_pass,
      family = "binomial")

summary(complete_ay_r)
```

Resulting in:

```
Call:
glm(formula = complete_pass ~ air_yards, family = "binomial",
    data = pbp_r_pass)

Coefficients:
              Estimate Std. Error z value Pr(>|z|)
(Intercept)  1.0719692  0.0080414  133.31   <2e-16 ***
air_yards   -0.0573223  0.0006244  -91.81   <2e-16 ***
---
Signif. codes:  0 '***' 0.001 '**' 0.01 '*' 0.05 '.' 0.1 ' ' 1

(Dispersion parameter for binomial family taken to be 1)

    Null deviance: 171714  on 131605  degrees of freedom
Residual deviance: 162145  on 131604  degrees of freedom
AIC: 162149

Number of Fisher Scoring iterations: 4
```

> Many of the tools and functions, such as summary(), that exist for working with OLS outputs in Python and lm outputs in R work on glm outputs as well.

Notice that the outputs from both Python and R are similar to the outputs in Chapters 3 and 4. For both of these models, as air_yards increases, the probability of completion decreases. You care about whether the coefficients differ from 0 to see if the coefficient is statistically important.

Some plots in this book, such as Figures 5-3 and 5-4, can take a while (several minutes or more) to complete. If you find yourself slowed by plotting times on a regular basis when working with your data, consider plotting summaries of data rather than raw data. For example, the binning used in "Exploratory Data Analysis" on page 58 is one approach. Other tools not covered in this book are *hexbin* plots, such as those created by the hexbin package in R (*https://oreil.ly/_QQDZ*) or hexbin() plot function in matplotlib (*https://oreil.ly/CdSQF*).

To help you see the results from logistic regressions, both Python and R have plotting tools. With Python, use regplot() from seaborn, but set the logistic option to True to create Figure 5-3 (to see why a linear regression is a bad idea for this model, use the default option of False and notice how the line goes above and below the data):

```python
## Python
sns.regplot(data=pbp_py_pass, x='air_yards', y='complete_pass',
            logistic=True,
            line_kws={'color': 'red'},
            scatter_kws={'alpha':0.05});
plt.show();
```

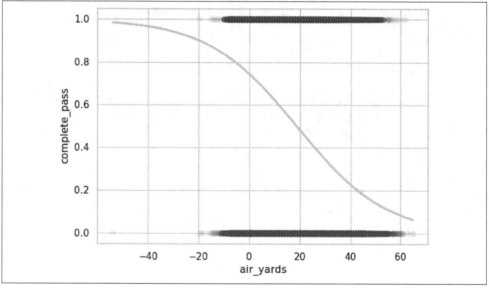

Figure 5-3. Pass completion as a function of air yards, plotted with a logistic curve in seaborn

In this plot, the curved line is the logistic function. The semitransparent points are the binary outcome for completed passes. Because of the large number of overlapping points, the logistic line is necessary to see any trends in the data.

Likewise, you can create a similar plot by using `ggplot2` in R, with jittering on the y-axis to help you see overlapping points (see Figure 5-4):

```R
## R
ggplot(data=pbp_r_pass,
       aes(x=air_yards, y=complete_pass)) +
  geom_jitter(height = 0.05, width = 0,
              alpha = 0.05) +
  stat_smooth(method = 'glm',
              method.args=list(family="binomial")) +
  theme_bw() +
  ylab("Completed pass (1 = yes, 0 = no)") +
  xlab("air yards")
```

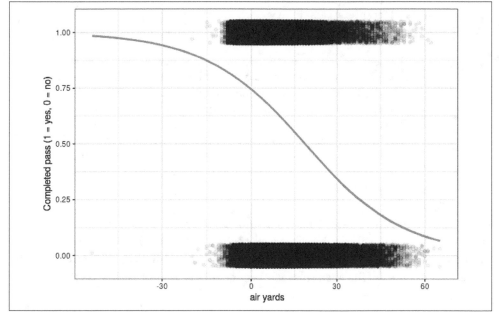

Figure 5-4. Pass completion as a function of air yards, plotted with a logistic curve in `ggplot2`

In this plot, the curved line is the logistic function. The semitransparent points are the binary outcome for completed passes. The points are jittered on the y-axis so that they are nonoverlapping. Because of the large number of overlapping points, the logistic line is necessary to see any trends in the data.

GLM Application to Completion Percentage

Using the results from "Building a GLM" on page 118, extract the expected completion percentage by appending the residuals to the play-by-play pass dataframe. With a linear model (or linear regression), the CPOE would simply be the residual because only one type exists. However, different types of residuals exist for GLMs, so you will calculate the residual manually (rather than extracting from the fit) to ensure that you know which type of residual you are using.

In Python, do this by using `predict()` to extract the predicted value from the model you previously fit and then subtract from the observed value to calculate the CPOE:

```
## Python
pbp_py_pass["exp_completion"] = \
  complete_ay_py.predict()

pbp_py_pass["cpoe"] = \
  pbp_py_pass["complete_pass"] - \
  pbp_py_pass["exp_completion"]
```

Because the GLM model occurs on a different scale than the observed data, various methods exist for calculating the residuals and predicted values. For example, the help file for the `predict()` function in R notes that multiple types of predictions exist. The default is on the scale of the linear predictors; the alternative response is on the scale of the response variable. Thus for a default binomial model, the default predictions are of log-odds (probabilities on logit scale) and `type = "response"` gives the predicted probabilities.

In R, take the `pbp_r_pass` dataframe and then create a new column by using `mutate()`. The new column is called `exp_completion` and gets values by extracting the predicted model fits via the `predict()` function with `type = "resp"` on the model you previously fit. Then subtract from `complete_pass` to calculate the CPOE:

```
## R
pbp_r_pass <-
  pbp_r_pass |>
  mutate(exp_completion = predict(complete_ay_r, type = "resp"),
         cpoe = complete_pass - exp_completion)
```

The code in this chapter is similar to the code in Chapters 3 and 4. If you need more details, refer to those chapters. Note, however, that GLMs differ from linear models in their output units and structure.

First, look at the leaders in CPOE since 2016, versus leaders in actual completion percentage. Recall that you're looking only at passes that have non-NA `air_yards` readings. Also, include only quarterbacks with 100 or more attempts. Filtering out the NA data helps you remove irrelevant plays. Filtering out to include only quarterbacks with 100 or more attempts avoids quarterbacks who had only a few plays and would likely be outliers.

In Python, calculate the mean CPOE and the mean completed pass percentage, and then sort by `compl`:

```Python
## Python
cpoe_py = \
  pbp_py_pass\
  .groupby(["season", "passer_id", "passer"])\
  .agg({"cpoe": ["count", "mean"],
        "complete_pass": ["mean"]})

cpoe_py.columns = \
  list(map('_'.join, cpoe_py.columns))
cpoe_py.reset_index(inplace=True)

cpoe_py = \
  cpoe_py\
   .rename(columns = {"cpoe_count": "n",
                      "cpoe_mean": "cpoe",
                      "complete_pass_mean": "compl"})\
   .query("n > 100")

print(
  cpoe_py\
   .sort_values("cpoe", ascending=False)
  )
```

Resulting in:

```
     season   passer_id     passer    n      cpoe     compl
299   2019   00-0020531    D.Brees   406  0.094099  0.756158
193   2018   00-0020531    D.Brees   566  0.086476  0.738516
467   2020   00-0033537   D.Watson   542  0.073453  0.704797
465   2020   00-0033357     T.Hill   121  0.072505  0.727273
22    2016   00-0026143     M.Ryan   631  0.068933  0.702060
..     ...      ...           ...    ...     ...       ...
91    2016   00-0033106     J.Goff   204 -0.108739  0.549020
526   2021   00-0027939   C.Newton   126 -0.109908  0.547619
112   2017   00-0025430  D.Stanton   159 -0.110229  0.496855
730   2022   00-0037327 S.Thompson   150 -0.116812  0.520000
163   2017   00-0031568    B.Petty   112 -0.151855  0.491071

[300 rows x 6 columns]
```

In R, calculate the mean CPOE and the mean completed pass percentage (compl), and then arrange by compl:

```R
## R
pbp_r_pass |>
  group_by(season, passer_id, passer) |>
  summarize(n = n(),
            cpoe = mean(cpoe, na.rm = TRUE),
            compl = mean(complete_pass, na.rm = TRUE),
            .groups = "drop") |>
  filter(n >= 100) |>
  arrange(-cpoe) |>
  print(n = 20)
```

Resulting in:

```
# A tibble: 300 × 6
   season passer_id   passer           n   cpoe compl
    <dbl> <chr>       <chr>        <int>  <dbl> <dbl>
 1   2019 00-0020531  D.Brees        406 0.0941 0.756
 2   2018 00-0020531  D.Brees        566 0.0865 0.739
 3   2020 00-0033537  D.Watson       542 0.0735 0.705
 4   2020 00-0033357  T.Hill         121 0.0725 0.727
 5   2016 00-0026143  M.Ryan         631 0.0689 0.702
 6   2019 00-0029701  R.Tannehill    343 0.0689 0.691
 7   2020 00-0023459  A.Rodgers      607 0.0618 0.705
 8   2017 00-0020531  D.Brees        606 0.0593 0.716
 9   2018 00-0026143  M.Ryan         607 0.0590 0.695
10   2021 00-0036442  J.Burrow       659 0.0564 0.703
11   2016 00-0020531  D.Brees        664 0.0548 0.708
12   2018 00-0032950  C.Wentz        399 0.0546 0.699
13   2018 00-0023682  R.Fitzpatrick  246 0.0541 0.667
14   2022 00-0030565  G.Smith        605 0.0539 0.701
15   2016 00-0027854  S.Bradford     551 0.0529 0.717
16   2018 00-0029604  K.Cousins      603 0.0525 0.705
17   2017 00-0031345  J.Garoppolo    176 0.0493 0.682
18   2022 00-0031503  J.Winston      113 0.0488 0.646
19   2021 00-0023459  A.Rodgers      556 0.0482 0.694
20   2020 00-0034857  J.Allen        692 0.0478 0.684
# i 280 more rows
```

Future Hall of Famer Drew Brees has not only some of the most accurate seasons in NFL history but also some of the most accurate seasons in NFL history even after adjusting for pass depth. In the results, Brees has four entries, all in the top 11. Cleveland Browns quarterback Deshaun Watson, who earned the richest fully guaranteed deal in NFL history at the time in 2022, scored incredibly well in 2020 at CPOE, while in 2016 Matt Ryan not only earned the league's MVP award, but also led the Atlanta Falcons to the Super Bowl while generating a 6.9% CPOE per pass attempt. Ryan's 2018 season also appears among the leaders.

In 2020, Aaron Rodgers was the league MVP. In 2021, Joe Burrow led the league in yards per attempt and completion percentage en route to leading the Bengals to their first Super Bowl appearance since 1988. Sam Bradford's 2016 season is still a top-five season of all time in terms of completion percentage, since passed a few times by Drew Brees and fellow Saints quarterback Taysom Hill (who also appeared in Chapter 4), but Bradford does not appear as a top player historically in CPOE, as discussed previously.

Pass depth is certainly not the only variable that matters in terms of completion percentage. Let's add a few more features to the model—namely, down (down), distance to go for a first down (ydstogo), distance to go to the end zone (yardline_100), pass location (pass_location), and whether the quarterback was hit (qb_hit; more on this later). The formula will also include an interaction between down and ydstogo.

First, change variables to factors in Python, select the columns you will use, and drop the NA values:

```
## Python
# remove missing data and format data
pbp_py_pass['down'] = pbp_py_pass['down'].astype(str)
pbp_py_pass['qb_hit'] = pbp_py_pass['qb_hit'].astype(str)

pbp_py_pass_no_miss = \
  pbp_py_pass[["passer", "passer_id", "season",
               "down", "qb_hit", "complete_pass",
               "ydstogo", "yardline_100",
               "air_yards",
               "pass_location"]]\
               .dropna(axis = 0)
```

Then build and fit the model in Python:

```
## Python
complete_more_py = \
  smf.glm(formula='complete_pass ~ down * ydstogo + ' +
                  'yardline_100 + air_yards + ' +
                  'pass_location + qb_hit',
          data=pbp_py_pass_no_miss,
          family=sm.families.Binomial())\
          .fit()
```

Next, extract the outputs and calculate the CPOE:

```
## Python
pbp_py_pass_no_miss["exp_completion"] = \
  complete_more_py.predict()

pbp_py_pass_no_miss["cpoe"] = \
  pbp_py_pass_no_miss["complete_pass"] - \
  pbp_py_pass_no_miss["exp_completion"]
```

Now, summarize the outputs, and reformat and rename the columns:

```Python
## Python
cpoe_py_more = \
    pbp_py_pass_no_miss\
    .groupby(["season", "passer_id", "passer"])\
    .agg({"cpoe": ["count", "mean"],
          "complete_pass": ["mean"],
          "exp_completion": ["mean"]})

cpoe_py_more.columns = \
    list(map('_'.join, cpoe_py_more.columns))
cpoe_py_more.reset_index(inplace=True)

cpoe_py_more = \
    cpoe_py_more\
    .rename(columns = {"cpoe_count": "n",
                       "cpoe_mean": "cpoe",
                       "complete_pass_mean": "compl",
                       "exp_completion_mean": "exp_completion"})\
    .query("n > 100")
```

Finally, print the top 20 entries (we encourage you to print more, as we print only a limited number of rows to save page space):

```Python
## Python
print(
    cpoe_py_more\
    .sort_values("cpoe", ascending=False)
    )
```

Resulting in:

```
     season  passer_id      passer   n      cpoe     compl  exp_completion
193    2018  00-0020531   D.Brees   566  0.088924  0.738516        0.649592
299    2019  00-0020531   D.Brees   406  0.087894  0.756158        0.668264
465    2020  00-0033357    T.Hill   121  0.082978  0.727273        0.644295
22     2016  00-0026143    M.Ryan   631  0.077565  0.702060        0.624495
467    2020  00-0033537  D.Watson   542  0.072763  0.704797        0.632034
..      ...        ...        ...  ...       ...       ...             ...
390    2019  00-0035040  D.Blough   174 -0.100327  0.540230        0.640557
506    2020  00-0036312   J.Luton   110 -0.107358  0.545455        0.652812
91     2016  00-0033106    J.Goff   204 -0.112375  0.549020        0.661395
526    2021  00-0027939  C.Newton   126 -0.123251  0.547619        0.670870
163    2017  00-0031568   B.Petty   112 -0.166726  0.491071        0.657798

[300 rows x 7 columns]
```

Likewise, in R, remove missing data and format the data:

```R
## R
pbp_r_pass_no_miss <-
    pbp_r_pass |>
    mutate(down = factor(down),
```

```R
                  qb_hit = factor(qb_hit)) |>
         filter(complete.cases(down, qb_hit, complete_pass,
                               ydstogo,yardline_100, air_yards,
                               pass_location, qb_hit))
```

Then run the model in R and save the outputs:

```R
## R
complete_more_r <-
  pbp_r_pass_no_miss  |>
  glm(formula = complete_pass ~ down * ydstogo + yardline_100 +
                  air_yards + pass_location + qb_hit,
      family = "binomial")
```

Next, calculate the CPOE:

```R
## R
pbp_r_pass_no_miss <-
  pbp_r_pass_no_miss |>
  mutate(exp_completion = predict(complete_more_r, type = "resp"),
         cpoe = complete_pass - exp_completion)
```

Summarize the data:

```R
## R
cpoe_more_r <-
  pbp_r_pass_no_miss |>
  group_by(season, passer_id, passer) |>
  summarize(n = n(),
            cpoe = mean(cpoe , na.rm = TRUE),
            compl = mean(complete_pass),
            exp_completion = mean(exp_completion),
            .groups = "drop") |>
  filter(n > 100)
```

Finally, print the top 20 entries (we encourage you to print more, as we print only a limited number of rows to save page space):

```R
## R
cpoe_more_r |>
  arrange(-cpoe) |>
  print(n = 20)
```

Resulting in:

```
# A tibble: 300 × 7
   season passer_id  passer           n   cpoe compl exp_completion
    <dbl> <chr>      <chr>        <int>  <dbl> <dbl>          <dbl>
 1   2018 00-0020531 D.Brees        566 0.0889 0.739          0.650
 2   2019 00-0020531 D.Brees        406 0.0879 0.756          0.668
 3   2020 00-0033357 T.Hill         121 0.0830 0.727          0.644
 4   2016 00-0026143 M.Ryan         631 0.0776 0.702          0.624
 5   2020 00-0033537 D.Watson       542 0.0728 0.705          0.632
 6   2019 00-0029701 R.Tannehill    343 0.0667 0.691          0.624
 7   2016 00-0027854 S.Bradford     551 0.0615 0.717          0.655
```

```
 8    2018 00-0023682 R.Fitzpatrick   246 0.0613 0.667      0.605
 9    2020 00-0023459 A.Rodgers       607 0.0612 0.705      0.644
10    2018 00-0026143 M.Ryan          607 0.0597 0.695      0.636
11    2018 00-0032950 C.Wentz         399 0.0582 0.699      0.641
12    2017 00-0020531 D.Brees         606 0.0574 0.716      0.659
13    2021 00-0036442 J.Burrow        659 0.0559 0.703      0.647
14    2016 00-0025708 M.Moore         122 0.0556 0.689      0.633
15    2022 00-0030565 G.Smith         605 0.0551 0.701      0.646
16    2021 00-0023459 A.Rodgers       556 0.0549 0.694      0.639
17    2017 00-0031345 J.Garoppolo     176 0.0541 0.682      0.628
18    2018 00-0033537 D.Watson        548 0.0539 0.682      0.629
19    2019 00-0029263 R.Wilson        573 0.0538 0.663      0.609
20    2018 00-0029604 K.Cousins       603 0.0533 0.705      0.652
# i 280 more rows
```

Notice that Brees's top seasons flip, with his 2018 season now the best in terms of CPOE, followed by 2019. This flip occurs because the model has slightly different estimates for the players based on the models having different features. Matt Ryan's 2016 MVP season eclipses Watson's 2020 campaign, while Sam Bradford enters back into the fray when we throw game conditions into the mix. Journeyman Ryan Fitzpatrick, who led the NFL in yards per attempt in 2018 for Tampa Bay while splitting time with Jameis Winston, joins the top group as well.

Is CPOE More Stable Than Completion Percentage?

Just as you did for running backs, it's important to determine whether CPOE is more stable than simple completion percentage. If it is, you can be sure that you're isolating the player's performance more so than his surrounding conditions. To this aim, let's dig into code.

First, calculate the lag between the current CPOE and the previous year's CPOE with Python:

```python
## Python
#  keep only the columns needed
cols_keep =\
    ["season", "passer_id", "passer",
     "cpoe", "compl", "exp_completion"]

# create current dataframe
cpoe_now_py =\
    cpoe_py_more[cols_keep].copy()

# create last-year's dataframe
cpoe_last_py =\
    cpoe_now_py[cols_keep].copy()

# rename columns
cpoe_last_py\
    .rename(columns = {'cpoe': 'cpoe_last',
```

```
                    'compl': 'compl_last',
                    'exp_completion': 'exp_completion_last'},
                   inplace=True)

# add 1 to season
cpoe_last_py["season"] += 1

# merge together
cpoe_lag_py =\
    cpoe_now_py\
    .merge(cpoe_last_py,
           how='inner',
           on=['passer_id', 'passer',
               'season'])
```

Then examine the correlation for pass completion:

```
## Python
cpoe_lag_py[['compl_last', 'compl']].corr()
```

Resulting in:

```
            compl_last      compl
compl_last    1.000000   0.445465
compl         0.445465   1.000000
```

Followed by CPOE:

```
## Python
cpoe_lag_py[['cpoe_last', 'cpoe']].corr()
```

Resulting in:

```
           cpoe_last       cpoe
cpoe_last   1.000000   0.464974
cpoe        0.464974   1.000000
```

You can also do these calculations in R:

```
## R
# create current dataframe
cpoe_now_r <-
    cpoe_more_r |>
    select(-n)

# create last-year's dataframe
# and add 1 to season
cpoe_last_r <-
    cpoe_more_r |>
    select(-n) |>
    mutate(season = season + 1) |>
    rename(cpoe_last = cpoe,
           compl_last = compl,
           exp_completion_last = exp_completion
           )
```

```
# merge together
cpoe_lag_r <-
    cpoe_now_r |>
    inner_join(cpoe_last_r,
               by = c("passer_id", "passer", "season")) |>
    ungroup()
```

Then select the two passing completion columns and examine the correlation:

```
## R
cpoe_lag_r |>
  select(compl_last, compl) |>
  cor(use="complete.obs")
```

Resulting in:

```
            compl_last      compl
compl_last   1.0000000 0.4454646
compl        0.4454646 1.0000000
```

Repeat with the CPOE columns:

```
## R
cpoe_lag_r |>
  select(cpoe_last, cpoe) |>
  cor(use="complete.obs")
```

Resulting in:

```
           cpoe_last       cpoe
cpoe_last  1.0000000 0.4649739
cpoe       0.4649739 1.0000000
```

It looks like CPOE is slightly more stable than pass completion! Thus, from a consistency perspective, you're slightly improving on the situation by building CPOE.

First, and most importantly: the features embedded in the expectation for completion percentage could be fundamental to the quarterback. Some quarterbacks, like Drew Brees, just throw shorter passes characteristically. Others take more hits. In fact, many, including Eric, have argued (*https://oreil.ly/Y1psK*) that taking hits is at least partially the quarterback's fault. Some teams throw a lot on early downs, which are easier passes to complete empirically, while others throw only on late downs. Quarterbacks don't switch teams that often, so even if the situation is necessarily inherent to the quarterback himself, the scheme in which they play may be stable.

Last, look at the stability of expected completions:

```
## R
cpoe_lag_r |>
  select(exp_completion_last, exp_completion) |>
  cor(use="complete.obs")
```

Resulting in:

```
                    exp_completion_last exp_completion
exp_completion_last           1.000000       0.473959
exp_completion                0.473959       1.000000
```

The most stable metric in this chapter is actually the average *expected completion percentage* for a quarterback.

A Question About Residual Metrics

The results of this chapter shed some light on issues that can arise in modeling football and sports in general. Trying to strip out context from a metric is rife with opportunities to make mistakes. The assumption that a player doesn't dictate the situation that they are embedded in on a given play is likely violated, and repeatedly.

For example, the NFL NGS version of CPOE includes receiver separation, which at first blush seems like an external factor to the quarterback: whether the receiver gets open is not the job of the quarterback. However, quarterbacks contribute to this in a few ways. First, the player they decide to throw to—the separation they choose from—is their choice. Second, quarterbacks can move defenders—and hence change the separation profile—with their eyes. Many will remember the no-look pass by Matthew Stafford in Super Bowl LVI.

Lastly, whether a quarterback actually passes the ball has some signal. As we've stated, Joe Burrow led the league in yards per attempt and completion percentage in 2021. He also led the NFL in sacks taken, with 51. Other quarterbacks escape pressure; some quarterbacks will run the ball, while others will throw it away. These alter expectations for reasons that are (at least partially) quarterback driven.

So, what should someone do? The answer here is the answer to a lot of questions, which is, it depends. If you're trying to determine the most accurate passer in the NFL, a single number might not necessarily be sufficient (no one metric is likely sufficient to answer this or any football question definitively).

If you're trying to predict player performance for the sake of player acquisition, fantasy football, or sports betting, it's probably OK to try to strip out context in an expectation and then apply the "over expected" analysis to a well-calibrated model. For example, if you're trying to predict Patrick Mahomes's completion percentage during the course of a season, you have to add the expected completion percentage given his circumstances and his CPOE. The latter is considered by many to be almost completely attributable to Mahomes, but the former also has some of his game embedded in it as well. To assume the two are completely independent will likely lead to errors.

As you gain access to more and better data, you will also gain the potential to reduce some of these modeling mistakes. This reduction requires diligence in modeling, however. That's what makes this process fun.

 We encourage you to refine your data skills before buying "better" data. Once you reach the limitations of the free data, you'll realize if and why you need better data. And, you'll be able to actually use that expensive data.

A Brief Primer on Odds Ratios

With a logistic regression, the coefficients may be understood in log-odds terminology as well. Most people do not understand log odds because odds are not commonly used in everyday life. Furthermore, the *odds* in odds ratios are different from betting odds.

 Odds ratios sometimes can help you understand logistic regression. Other times, they can lead you astray—far astray.

For example, with odds ratios, if you expect three passes to be completed for every two passes that were incomplete, then the odds ratio would be 3-to-2. The 3-to-2 odds may also be written in decimal form as an odds ratio of 1.5 (because $\frac{3}{2} = 1.5$), with an implied 1 for 1.5-to-1.

Odds ratios may be calculated by taking the exponential of the logistic regression coefficients (the e^x or exp() function on many calculators). For example, the broom package has a tidy() function that readily allows odds ratios to be calculated and displayed:

```
## R
complete_ay_r |>
  tidy(exponentiate = TRUE, conf.int = TRUE)
```

Resulting in:

```
# A tibble: 2 × 7
  term        estimate std.error statistic p.value conf.low conf.high
  <chr>          <dbl>     <dbl>     <dbl>   <dbl>    <dbl>     <dbl>
1 (Intercept)     2.92   0.00804     133.        0     2.88      2.97
2 air_yards       0.944  0.000624    -91.8       0     0.943     0.945
```

On the odds-ratio scale, you care if a value differs from 1 because odds of 1:1 implies that the predictor does not change the outcome of an event or the coefficient has no effect on the model's prediction. This intercept now tells you that a pass with 0 yards to go will be completed with odds of `2.92`. However, each additional yard decreases the odds ratio by odds of `0.94`.

To see how much each additional yard decreases the odds ratio, multiply the intercept and the `air_yards` coefficient. First, use the `air_yards` coefficient multiplied by itself for each additional yard to go (such as `air_yards` × `air_yards`) or, more generically, take `air_yards` to the yards-to-go power (for example air_yards^2 for 2 yards or air_yards^9 for 9 yards). For example, looking at Figures 5-3 of 5-4, you can see that passes with air yards approximately greater than 20 yards have a less than 50% chance of completion. If you multiply 2.92 by 0.94 to the 20th power (2.92×20^{20}), you can see that the probability of completing a pass with 2 yards to go is 85, which is slightly less than 50% and agrees with Figures 5-3 and 5-4.

To help better understand odds ratios, we will show you how to calculate them in R; we use R because the language has nicer tools for working with `glm()` outputs, and following the calculations is more important than being able to do them. First, calculate the completion percentage for all data in the play-by-play pass dataset. Next, calculate odds by taking the completion percentage and dividing by 1 minus the completion percentage. Then take the natural log to calculate the log odds:

```
## R
pbp_r_pass |>
  summarize(comp_pct = mean(complete_pass)) |>
  mutate(odds = comp_pct / (1 - comp_pct),
         log_odds = log(odds))
```

Resulting in:

```
# A tibble: 1 × 3
  comp_pct  odds log_odds
     <dbl> <dbl>    <dbl>
1    0.642  1.79    0.583
```

Next, compare this output to the logistic regression output for a logistic regression with only a global intercept (that is, an average across all observations). First, build the model with a global intercept (`complete_pass ~ 1`). Look at the `tidy()` coefficients for the raw and exponentiated outputs:

```
## R
complete_global_r <-
  glm(complete_pass ~ 1,
      data = pbp_r_pass,
      family = "binomial")

complete_global_r |>
  tidy()
```

Resulting in:

```
# A tibble: 1 × 5
  term        estimate std.error statistic p.value
  <chr>          <dbl>     <dbl>     <dbl>   <dbl>
1 (Intercept)    0.583   0.00575      101.       0
```

```
complete_global_r |>
  tidy(exponentiate = TRUE)
```

Resulting in:

```
# A tibble: 1 × 5
  term        estimate std.error statistic p.value
  <chr>          <dbl>     <dbl>     <dbl>   <dbl>
1 (Intercept)     1.79   0.00575      101.       0
```

Compare the outputs to the numbers you previously calculated using the percentages. Now, hopefully, you will never need to calculate odds ratios by hand again!

Data Science Tools Used in This Chapter

This chapter covered the following topics:

- Fitting a logistic regression in Python and R by using `glm()`
- Understanding and reading the coefficients from a logistic regression, including odds ratios
- Reapplying data-wrangling tools you learned in previous chapters

Exercises

1. Repeat this analysis without `qb_hit` as one of the features. How does it change the leaderboard? What can you take from this?
2. What other features could be added to the logistic regression? How does it affect the stability results in this chapter?
3. Try this analysis for receivers. Does anything interesting emerge?
4. Try this analysis for defensive positions. Does anything interesting emerge?

Suggested Readings

The resources suggested in "Suggested Readings" on pages 76 and 112 will help you understand generalized linear models. Some other readings include the following:

- A PFF article by Eric about quarterbacks, "Quarterbacks in Control: A PFF Data Study of Who Controls Pressure Rates" (*https://oreil.ly/lhIJu*).

- *Beyond Multiple Linear Regression: Applied Generalized Linear Models and Multi-level Models in R* by Paul Roback and Julie Legler (CRC Press, 2021). As the title suggests, this book goes beyond linear regression and does a nice job of teaching generalized linear models, including the model's important assumptions.
- *Bayesian Data Analysis*, 3rd edition, by Andrew Gelman et al. (CRC Press, 2013). This book is a classic for advanced modeling skills but also requires a solid understanding of linear algebra.

## CHAPTER 6

Using Data Science for Sports Betting: Poisson Regression and Passing Touchdowns

Much progress has been made in the arena of sports betting in the United States specifically and the world broadly. While Eric was at the Super Bowl in February 2023, almost every single television and radio show in sight was sponsored by some sort of gaming entity. Just five years earlier, sports betting was a taboo topic, discussed only by the fringes, and not legal in any state other than Nevada. That all changed in the spring of 2018, when the Professional and Amateur Sports Protection Act (PASPA) was repealed by the US Supreme Court, allowing states to determine if and how they would legalize sports betting within their borders.

Sports betting started slowly legalizing throughout the US, with New Jersey and Pennsylvania early adopters, before spreading west to spots including Illinois and Arizona. Almost two-thirds of states now have some form of legalized wagering, which has caused a gold rush in offering new and varied products for gamblers—both recreational and professional.

The betting markets are the single best predictor of what is going to occur on the football field any given weekend for one reason: the wisdom of the crowds. This topic is covered in *The Wisdom of the Crowds* by James Surowiecki (Doubleday, 2004). Market-making *books* (in gambling, *books* are companies that take bets, and *bookies* are individuals who do the same thing) like Pinnacle, Betcris, and Circa Sports have oddsmakers that *set the line* by using a process called *origination* to create the initial price for a game.

Early in the week, bettors—both recreational and professional alike—will stake their opinions through making wagers. These wagers are allowed to increase as the week (and information like weather and injuries) progresses. The *closing line*, in theory, contains all the opinions, expressed through wagers, of all bettors who have enough influence to move the line into place.

Because markets are assumed to tend toward efficiency, the final prices on a game (described in the next section) are the most accurate (public) predictions available to us. To beat the betting markets, you need to have an *edge*, which is information that gives an advantage not available to other bettors.

An edge can generally come from two sources: better data than the market, or a better way of synthesizing data than the market. The former is usually the product of obtaining information (like injuries) more quickly than the rest of the market, or collecting *longitudinal* data (detailed observations through time, in this game-level data) that no one else bothers to collect (as PFF did in the early days). The latter is generally the approach of most bettors, who use statistical techniques to process data and produce models to set their own prices and bet the discrepancies between their price (an *internal*) and that of the market. That will be the main theme of this chapter.

The Main Markets in Football

In American football, the three main markets have long been the spread, the total, and the moneyline. The *spread*, which is the most popular market, is pretty easy to understand: it is a point value meant to split outcomes in half over a large sample of games. Let's consider an example. The Washington Commanders are—in a theoretical world where they can play an infinite number of games under the same conditions against the New York Giants—an average of four points better on the field. Oddsmakers would therefore make the spread between the Commanders and the Giants four points. With this example, five outcomes can occur for betting:

- A person bets for the Commanders to win. The Commanders win by five or more points, and the person wins their bet.

- A person bets for the Commanders to win. The Commanders lose outright or do not win by five or more points, and the person loses their bet.

- A person bets for the Giants to win. The Giants either win in the game outright or lose by three or fewer points, and the person wins their bet.

- A person bets for the Giants to win. The Giants lose outright, and the person loses their bet.

- A person bets for either team to win and the game *lands on* (the final score is) a four-point differential in favor of the Commanders. This game would be deemed a *push*, with the bettor getting their money back.

It's expected that a point-spread bettor without an advantage over the sportsbook would win about 50% of their bets. This 50% comes from the understanding of probability, where the spread (theoretically) captures all information about the system. For example, betting on a coin being heads will be correct half the time and incorrect half the time over a long period. For the sportsbook to earn money off this player, it charges a *vigorish*, or *vig*, on each bet. For point-spread bets, this is typically 10 cents on the dollar.

Hence, if you wanted to win $100 betting Washington (–4), you would *lay* $110 to win $100. Such a bet requires a 52.38% success rate (computed by taking 110 / (110 + 100)) on nonpushed bets over the long haul to break even. Hence, to win at point-spread betting, you need roughly a 2.5-point edge (52.38%–50% is roughly 2.5%) over the sportsbook, which given the fact that over 99% of sports bettors lose money, is a lot harder than it sounds.

The old expression "The house always wins" occurs because the house plays the long game, has built-in advantages, and wins at least slightly more than it loses. With a roulette table, for example, the house advantage comes from the inclusion of the green 0 number or numbers (so you have less than a 50-50 chance of getting red or black). In the case of sports betting, the house gets the vig, which can almost guarantee it a profit unless its odds are consistently and systematically wrong.

To bet a *total*, you simply bet on whether the sum of the two teams' points goes over or under a specified amount. For example, say the Commanders and Giants have a market total of 43.5 points (–110); to bet under, you'd have to lay $110 to win $100 and hope that the sum of the Commanders' and Giants' points was 43 or less. Forty-four or more points would result in a loss of your initial $110 stake. No pushes are possible when the spread or total has a 0.5 tacked onto it. Some bettors specialize in totals, both full-game totals and totals that are applicable for only certain segments of the game (such as first half or first quarter).

The last of the traditional bets in American football is the *moneyline bet*. Essentially, you're betting on a team to win the game straight up. Since a game is rarely a true 50-50 (pick'em) proposition, to bet a moneyline, you either have to lay a lot of money to win a little money (when a team is a *favorite*), or you get to bet a little money to win a lot of money (when a team is an *underdog*). For example, if the Commanders are considered 60% likely to win against the Giants, the Commanders would have a moneyline price (using North American odds, other countries use decimal odds) of –150; the bettor needs to lay $150 to win $100. The decimal odds for this bet are (100 + 150) / 150 = 1.67, which is the ratio of the total return to the investment. –150 is arrived at partially through convention—namely, the minus sign in front of the odds for a favorite, and through the computation of $100 \times \frac{0.6}{(1 - 0.6)} = 150$. The decimal odds here are $(100 + 150) / 100 = 2.5$.

The Giants would have a price of +150, meaning that a successful bet of $100 would pay out $150 in addition to the original bet. This is arrived at in the reciprocal way: $100 \times \frac{1 - 0.4}{0.4} = 150$, with the convention that the plus sign goes in front of the price for the underdog.

 The book takes some vigorish in moneyline bets, too, so instead of Washington being –150 and New York being +150, you might see something like Washington being closer to –160 and New York being closer to +140 (vigs vary by book). The daylight between the absolute values of –160 and +140 is present in all but rare promotional markets for big games like the Super Bowl.

Application of Poisson Regression: Prop Markets

A model worth its salt in the three main football-betting markets using regression is beyond the scope of this book, as they require ratings for each team's offense, defense, and special teams and require adjustments for things like weather and injuries. They are also the markets that attract the highest number of bettors and the most betting *handle* (or total amount of money placed by bettor), which makes them the most efficient betting markets in the US and among the most efficient betting markets in the world.

Since the overturning of PASPA, however, sportsbook operators have rushed to create alternatives for bettors who don't want to swim in the deep seas of the spreads, totals, and moneylines of the football-betting markets. As a result, we see the proliferation of *proposition* (or *prop*) markets. Historically reserved for big events like the Super Bowl, now all NFL games and most college football games offer bettors the opportunity to bet on all kinds of events (or *props*): Who will score the first touchdown? How many interceptions will Patrick Mahomes have? How many receptions will Tyreek Hill have? Given the sheer volume of available wagers here, it's much, much more difficult for the sportsbook to get each of these prices right, and hence a bigger opportunity for bettors to exist in these prop markets.

In this chapter, you'll examine the touchdown pass market for NFL quarterbacks. Generally speaking, the quarterback starting a game will have a prop market of over/under 0.5 touchdown passes, over/under 1.5 touchdown passes, and for the very best quarterbacks, over/under 2.5 touchdown passes. The number of touchdown passes is called the *index* in this case. Since the number of touchdown passes a quarterback throws in a game is so discrete, the most important aspect of the prop offering is the price on over and under, which is how the markets create a continuum of offerings in response to bettors' opinions.

Thus, for the most popular index, over/under 1.5 touchdown passes, one player might have a price of –140 (lay $140 to win $100) on the over, while another player may have a price of +140 (bet $100 to win $140) on over the same number of touchdown passes. The former is a *favorite* to go over 1.5 touchdown passes, while the latter is an underdog to do so. The way these are determined, and whether you should bet them, is determined largely by analytics.

The Poisson Distribution

To create or bet a proposition bet or participate in any betting market, you have to be able to estimate the likelihood, or *probability*, of events happening. In the canonical example in this chapter, this is the number of touchdown passes thrown by a particular quarterback in a given game.

The simplest way to try to specify these probabilities is to empirically look at the frequencies of each outcome: zero touchdown passes, one touchdown pass, two touchdown passes, and so on. Let's look at NFL quarterbacks with at least 10 passing plays in a given week from 2016 to 2022 to see the frequencies of various touchdown-pass outcomes. We use the 10 passing plays threshold as a proxy for being the team's starter, which isn't perfect, but will do for now. Generally speaking, passing touchdown props are offered only for the starters in a given game. You will also use the same filter as in Chapter 5 to remove nonpassing plays. First, load the data in Python:

```
## Python
import pandas as pd
import numpy as np
import nfl_data_py as nfl
import statsmodels.formula.api as smf
import statsmodels.api as sm
import matplotlib.pyplot as plt
import seaborn as sns
from scipy.stats import poisson

seasons = range(2016, 2022 + 1)
pbp_py =\
    nfl.import_pbp_data(seasons)

pbp_py_pass = \
  pbp_py.\
    query('passer_id.notnull()')\
    .reset_index()
```

Or load the data in R:

```r
## R
library(nflfastR)
library(tidyverse)

pbp_r <-
    load_pbp(2016:2022)

pbp_r_pass <-
  pbp_r |>
  filter(!is.na(passer_id))
```

Then replace NULL or NA values with 0 for `pass_touchdown`. Python also requires plays without a `passer_id` and `passer` to be set to `none` so that the data will be summarized correctly.

Next, aggregate by `season`, `week`, `passer_id`, and `passer` to calculate the number of passes per week and the number of touchdown passes per week. Then, filter to exclude players with fewer than 10 plays as a passer for each week. Next, calculate the number of touchdown passes per quarterback per week.

Lastly, save the `total_line` because you will use this later. This is just nflfastR's name for the market for total points scored, which we discussed earlier in this chapter. We assume that games with different totals will have different opportunities for touchdown passes (e.g., higher totals will have more touchdown passes, on average). The `total_line` is the same throughout a game, so you need to use a function so that Python or R can aggregate a value for the game. A function like `mean()` or `max()` will give you the value for the game, and we used `mean()`. Use this code in Python:

```python
## Python
pbp_py_pass\
    .loc[pbp_py_pass.pass_touchdown.isnull(), "pass_touchdown"] = 0

pbp_py_pass\
    .loc[pbp_py_pass.passer.isnull(), "passer"] = 'none'

pbp_py_pass\
    .loc[pbp_py_pass.passer_id.isnull(), "passer_id"] = 'none'

pbp_py_pass_td_y = \
    pbp_py_pass\
    .groupby(["season", "week", "passer_id", "passer"])\
    .agg({"pass_touchdown": ["sum"],
          "total_line": ["count", "mean"]})

pbp_py_pass_td_y.columns =\
    list(map("_".join, pbp_py_pass_td_y.columns))

pbp_py_pass_td_y.reset_index(inplace=True)
```

```python
pbp_py_pass_td_y\
    .rename(columns={
        "pass_touchdown_sum": "pass_td_y",
        "total_line_mean": "total_line",
        "total_line_count": "n_passes"
    },
    inplace=True
)

pbp_py_pass_td_y =\
    pbp_py_pass_td_y\
    .query("n_passes >= 10")

pbp_py_pass_td_y\
    .groupby("pass_td_y")\
    .agg({"n_passes": "count"})
```

Resulting in:

```
          n_passes
pass_td_y
0.0            902
1.0           1286
2.0           1050
3.0            506
4.0            186
5.0             31
6.0              4
```

Or use this code in R:

```r
## R
pbp_r_pass_td_y <-
    pbp_r_pass |>
    mutate(
        pass_touchdown = ifelse(is.na(pass_touchdown), 0,
                                pass_touchdown)) |>
    group_by(season, week, passer_id, passer) |>
    summarize(
        n_passes = n(),
        pass_td_y = sum(pass_touchdown),
        total_line = mean(total_line)
    ) |>
    filter(n_passes >= 10)

pbp_r_pass_td_y |>
    group_by(pass_td_y) |>
    summarize(n = n())
```

Resulting in:

```
# A tibble: 7 × 2
  pass_td_y     n
      <dbl> <int>
1         0   902
2         1  1286
3         2  1050
4         3   506
5         4   186
6         5    31
7         6     4
```

 You are able to group by `season` and `week` because each team has only one game per week. You group by `passer_id` and `passer` because `passer_id` is unique (some quarterbacks might have the same name, or at least first initial and last name). You include `passer` because this helps to better understand the data. When using groupings like this on new data, think through how to create unique groups for your specific needs.

Now you can see why the most popular index is 1.5, since the meat of the empirical distribution is centered at around one touchdown pass, with players with at least ten pass attempts more likely to throw two or more touchdown passes than they are to throw zero. The mean of the distribution is 1.48 touchdown passes, as you can see here in Python:

```
## Python
pbp_py_pass_td_y\
    .describe()
```

Resulting in:

	season	week	pass_td_y	n_passes	total_line
count	3965.000000	3965.000000	3965.000000	3965.000000	3965.000000
mean	2019.048928	9.620177	1.469609	38.798487	45.770618
std	2.008968	5.391064	1.164085	10.620958	4.409124
min	2016.000000	1.000000	0.000000	10.000000	32.000000
25%	2017.000000	5.000000	1.000000	32.000000	42.500000
50%	2019.000000	10.000000	1.000000	39.000000	45.500000
75%	2021.000000	14.000000	2.000000	46.000000	48.500000
max	2022.000000	22.000000	6.000000	84.000000	63.500000

Or in R:

```
pbp_r_pass_td_y |>
    ungroup() |>
    select(-passer, -passer_id) |>
    summary()
```

Resulting in:

```
      season            week          n_passes         pass_td_y        total_line
 Min.   :2016     Min.   : 1.00    Min.   :10.0    Min.   :0.00    Min.   :32.00
 1st Qu.:2017     1st Qu.: 5.00    1st Qu.:32.0    1st Qu.:1.00    1st Qu.:42.50
 Median :2019     Median :10.00    Median :39.0    Median :1.00    Median :45.50
 Mean   :2019     Mean   : 9.62    Mean   :38.8    Mean   :1.47    Mean   :45.77
 3rd Qu.:2021     3rd Qu.:14.00    3rd Qu.:46.0    3rd Qu.:2.00    3rd Qu.:48.50
 Max.   :2022     Max.   :22.00    Max.   :84.0    Max.   :6.00    Max.   :63.50
```

Counts of values are a good place to start, but sometimes you'll need something more. In general, relying solely on these counts to make inferences and predictions has numerous issues. The most important issue that arises is one of generalization. This is where probability distributions come in handy.

Touchdown passes aren't the only prop market in which you're going to want to make bets; things like interceptions, sacks, and other low-frequency markets may all have similar quantitative features, and it would benefit you to have a small set of tools in the toolbox from which to work. Furthermore, other markets like passing yards, for which there are tenfold discrete outcomes, can often have more potential outcomes than outcomes that have occurred in the history of a league and, very likely, the history of a player. A general framework is evidently necessary here.

This is where probability distributions come in handy. A *probability distribution* is a mathematical object that assigns to each possible outcome a value between 0 and 1, called a *probability*. For discrete outcomes, like touchdown passes in a game, this is pretty easy to understand, and while it might require a formula to compute for each outcome, you can generally get the answer to the question "what is the probability that $X = 0$?" For outcomes that are continuous, like heights, it's a bit more of a chore and requires tools from calculus. We will stick with using discrete probability distributions in this book.

One of the most popular discrete probability distributions is the *Poisson distribution*. This distribution defines the probability of obtaining the integer (that is, the discrete value) x (such as $x = 0, 1, 2, 3, ...$) as the value $\frac{e^\lambda \lambda^x}{x!}$. In this equation, the Greek letter λ (lambda) is the average value of the population, and ! is the factorial function. The Poisson distribution models the likelihood of a given number of events occurring in a fixed interval of time or space.

The definition of a factorial is $n! = n \times (n - 1) \times (n - 2) \times (n - 3)...$ $\times 2 \times 1$ and $0! = 1$. You might also remember them from math class for their use with permutations. For example, how many ways can we arrange three letters (a, b, and c)? 3! = 6, or aba, acb, bac, bcb, cab, and cba.

Critical assumptions of the Poisson distribution are as follows:

- The events occur with equal probability.
- The events are independent of the time since the last event.

These assumptions are not exactly satisfied in football, as a team that scores one touchdown in a game may or may not be likely to have "figured out" the defense on the other side of the field, but it's at least a distribution to consider in modeling touchdown passes in a game by a quarterback.

 Both Python and R have powerful tools for working with statistical distributions. We only touch on these topics in this book. We have found books such as Benjamin M. Bolker's *Ecological Models and Data in R* (Princeton University Press, 2008) to be great resources on applied statistical distributions and their applications.

To see if a Poisson is reasonable, let's look at a bar graph of the frequencies and compare this with the Poisson distribution of the same mean, λ. Do this using this Python code to create Figure 6-1:

```Python
## Python
pass_td_y_mean_py =\
    pbp_py_pass_td_y\
    .pass_td_y\
    .mean()

plot_pos_py =\
    pd.DataFrame(
        {"x": range(0, 7),
        "expected": [poisson.pmf(x, pass_td_y_mean_py) for x in range(0, 7)]
        }
    )

sns.histplot(pbp_py_pass_td_y["pass_td_y"], stat="probability");
plt.plot(plot_pos_py.x, plot_pos_py.expected);
plt.show();
```

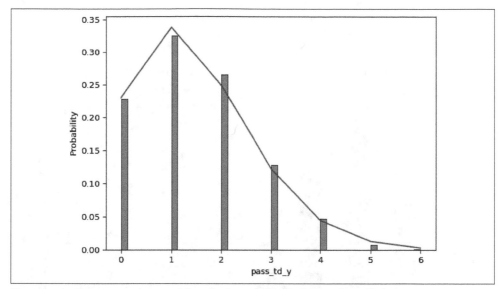

Figure 6-1. Histogram (vertical bars) of normalized observed touchdowns per game per quarterback with at least 10 games (plotted with seaborn)

For this histogram, the term *normalized* refers to all the bars summing to 1. The line shows the theoretical expected values from the Poisson distribution.

Or, use this R code to create Figure 6-2:

```
## R
pass_td_y_mean_r <-
    pbp_r_pass_td_y |>
    pull(pass_td_y) |>
    mean()

plot_pos_r <-
    tibble(x = seq(0, 7)) |>
    mutate(expected = dpois(
        x = x,
        lambda = pass_td_y_mean_r
    ))

ggplot() +
    geom_histogram(
        data = pbp_r_pass_td_y,
        aes(
            x = pass_td_y,
            y = after_stat(count / sum(count))
        ),
        binwidth = 0.5
    ) +
    geom_line(
```

```
        data = plot_pos_r, aes(x = x, y = expected),
        color = "red", linewidth = 1
) +
theme_bw() +
xlab("Touchdown passes per player per game for 2016 to 2022") +
ylab("Probability")
```

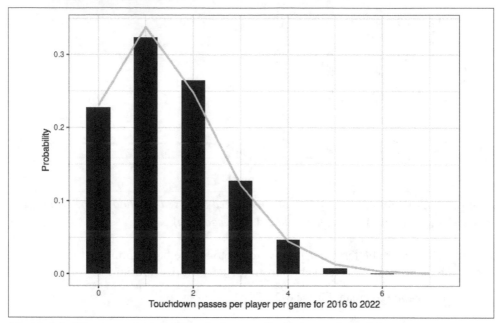

Figure 6-2. Histogram (vertical bars) of normalized observed touchdowns per game per quarterback with at least 10 games (plotted with `ggplot2`)

For this histogram, the *normalized* means that all the bars sum to 1. The line shows the theoretical expected values from the Poisson distribution. The Poisson distribution seems to slightly overestimate the likelihood of one touchdown pass, and as such, slightly underestimate the likelihood of zero, two, or more touchdown passes in a game.

Although not large, such discrepancies can be the difference between winning and losing in sports betting, so if you want to wager on your opinions, some adjustments need to be made. Alternatively, you could use a different distribution entirely (such as a negative binomial distribution or quasi-Poisson that accounts for over dispersion). But for the sake of this book, we're going to assume that a Poisson distribution is sufficient to handicap touchdown pass probabilities in this chapter.

Individual Player Markets and Modeling

Each quarterback, and the player's opponent, varies in quality, and hence each player will have a different market each game regarding touchdown passes. For example, in Super Bowl LVII between the Kansas City Chiefs and Philadelphia Eagles, Patrick Mahomes had a touch-down pass prop of 2.5 touchdown passes (per DraftKings Sportsbook), with the over priced at +150 (bet $100 to win $150) and the under priced at –185 (bet $185 to win $100).

As we've discussed, these prices reflect probabilities that you are betting against. In the case of the over, since the price is > 0, the formula for computing the break-even probability is 100 / (100 + 150) = 0.4. In other words, to bet the over, you have to be confident that Mahomes has better than a 40% chance to throw three or more touchdown passes against the Eagles defense, which ranked number one in the NFL during the 2022 season.

For the under, the break-even probability is 185 / (100 + 185) = 0.649, and hence you have to be more than 64.9% confident that Mahomes will throw for two or fewer touchdowns to make a bet on the under. Notice that these probabilities add to 104.9%, with the 4.9% representing the house edge, or *hold*, which is created by the book charging the vig discussed previously.

To specify these probabilities for Mahomes, you could simply go through the history of games Mahomes has played and look at the proportion of games with two or fewer touchdowns and three or more touchdowns to compare. This is faulty for a few reasons, the first of which is that it doesn't consider exogenous factors like opponent strength, weather, changes to supporting cast, or similar factors. It also doesn't factor changes to league-wide environments, like what happened during COVID-19, where the lack of crowd noise significantly helped offenses.

Now, you can incorporate this into a model in manifold ways, and anyone who is betting for real should consider as many factors as is reasonable. Here, you will use the aforementioned *total* for the game—the number of points expected by the betting markets to be scored. This factors in defensive strength, pace, and weather together as one number. In addition to this number, you will use as a feature the mean number of touchdown passes by the quarterback of interest over the previous two seasons.

The for-loop is a powerful tool in programming. When we are starting to build for loops, and especially nested for loops, we will start and simply print the indexes. For example, we would use code in Python and then fill in details for each index:

```Python
## Python
for season_idx in range(2017, 2022 + 1):
    print(season_idx)
    for week_idx in range(1, 22 + 1):
        print(week_idx)
```

We do this for two reasons. First, this makes sure the indexing is working. Second, we now have season_idx and week_idx in memory. If we can get our code working for these two index examples, there is a good chance our code will work for the rest of the index values in the for loop.

This is what you're going to call *x* in your training of the model. With Python, use this code:

```Python
## Python
# pass_ty_d greater than or equal to 10 per week
pbp_py_pass_td_y_geq10 =\
    pbp_py_pass_td_y.query("n_passes >= 10")

# take the average touchdown passes for each QB for the previous season
# and current season up to the current game
x_py = pd.DataFrame()
for season_idx in range(2017, 2022 + 1):
    for week_idx in range(1, 22 + 1):
        week_calc_py = (
            pbp_py_pass_td_y_geq10\
                .query("(season == " +
                        str(season_idx - 1) +
                        ") |" +
                        "(season == " +
                        str(season_idx) +
                        "&" +
                        "week < " +
                        str(week_idx) +
                        ")")\
            .groupby(["passer_id", "passer"])\
            .agg({"pass_td_y": ["count", "mean"]})
        )
        week_calc_py.columns =\
            list(map("_".join, week_calc_py.columns))
        week_calc_py.reset_index(inplace=True)
        week_calc_py\
            .rename(columns={
                "pass_td_y_count": "n_games",
                "pass_td_y_mean": "pass_td_rate"},
            inplace=True)
```

```
week_calc_py["season"] = season_idx
week_calc_py["week"] = week_idx
x_py = pd.concat([x_py, week_calc_py])
```

 Nested loops can quickly escalate computational times and decrease code readability. If you find yourself using many nested loops, consider learning other coding methods such as vectorization. Here, we use loops because loops are easier to understand and because computer performance is not important.

Or with R, use this code:

```
## R
# pass_ty_d greater than or equal to 10 per week
pbp_r_pass_td_y_geq10 <-
    pbp_r_pass_td_y |>
    filter(n_passes >= 10)

# take the average touchdown passes for each QB for the previous season
# and current season up to the current game
x_r <- tibble()

for (season_idx in seq(2017, 2022)) {
    for (week_idx in seq(1, 22)) {
        week_calc_r <-
            pbp_r_pass_td_y_geq10 |>
            filter((season == (season_idx - 1)) |
                (season == season_idx & week < week_idx)) |>
            group_by(passer_id, passer) |>
            summarize(
                n_games = n(),
                pass_td_rate = mean(pass_td_y),
                .groups = "keep"
            ) |>
            mutate(season = season_idx, week = week_idx)

        x_r <- bind_rows(x_r, week_calc_r)
    }
}
```

 By historic convention, many people use i, j, and k for the indexes in for loops such as for i in …. Richard prefers to use longer terms like season_idx or week_idx for three reasons. First, the words are more descriptive and help him see what is going on in the code. Second, the words are easier to search for with the Find tools. Third, the words are less likely to be repeated elsewhere in the code.

Notice here for every player going into each week, you have their average number of touchdown passes.

 We use for loops in the book because they are conceptually simple to use and understand. Other tools exist such as map() in Python or *apply* functions in R such as lappy(), or apply(). These functions are quicker, easier for advanced users to understand and read, and often less error prone. However, we wrote a book about introductory football analytics, not advanced data science programming. Hence, we usually stick to for loops for this book. See resources such as Chapter 9 (*https://adv-r.hadley.nz/functionals.html*) in Hadley Wickham's *Advanced R*, 2nd edition (CRC Press, 2019), which describe these methods and why to use them.

Let's look at Patrick Mahomes going into Super Bowl LVII in Python:

```
## Python
x_py.query('passer == "P.Mahomes"').tail()
```

Resulting in:

```
     passer_id      passer   n_games   pass_td_rate   season   week
39   00-0033873   P.Mahomes       36       2.444444     2022     18
40   00-0033873   P.Mahomes       37       2.405405     2022     19
40   00-0033873   P.Mahomes       37       2.405405     2022     20
40   00-0033873   P.Mahomes       38       2.394737     2022     21
40   00-0033873   P.Mahomes       39       2.384615     2022     22
```

Or, let's look in R:

```
## R
x_r |>
    filter(passer == "P.Mahomes") |>
    tail()
```

Resulting in:

```
# A tibble: 6 × 6
  passer_id   passer    n_games  pass_td_rate  season   week
  <chr>       <chr>       <int>        <dbl>   <int>  <int>
1 00-0033873  P.Mahomes      35         2.43    2022     17
2 00-0033873  P.Mahomes      36         2.44    2022     18
3 00-0033873  P.Mahomes      37         2.41    2022     19
4 00-0033873  P.Mahomes      37         2.41    2022     20
5 00-0033873  P.Mahomes      38         2.39    2022     21
6 00-0033873  P.Mahomes      39         2.38    2022     22
```

Looks like the books set a decent number. Mahomes's average prior to that game using data from 2021 and 2022 up to week 22 was 2.38 touchdown passes per game. You now have to create your response variable, which is simply the dataframe `pbp_pass_td_y_geq10` created with the previous code along with the added game total. In Python, use a `merge()` function:

```
## Python
pbp_py_pass_td_y_geq10 =\
    pbp_py_pass_td_y_geq10.query("season != 2016")\
    .merge(x_py,
           on=["season", "week", "passer_id", "passer"],
           how="inner")
```

In R, use an `inner_join()` function (Appendix C provides more details on joins):

```
### R
pbp_r_pass_td_y_geq10 <-
    pbp_r_pass_td_y_geq10 |>
    inner_join(x_r,
        by = c(
            "season", "week",
            "passer_id", "passer"
        )
    )
```

You've now merged the datasets together to get your training dataset for the model. Before you model that data, quickly peek at it by using `ggplot2` in R. First, plot passing touchdowns in each game for each passer by using a line (using the `passer_id` column rather than `passer`). In the plot, `facet` by season and add a meaningful caption. Save this as _weekly_passing_id_r_plot_ and look at Figure 6-3:

```
## R
weekly_passing_id_r_plot <-
    pbp_r_pass_td_y_geq10 |>
    ggplot(aes(x = week, y = pass_td_y, group = passer_id)) +
    geom_line(alpha = 0.25) +
    facet_wrap(vars(season), nrow = 3) +
    theme_bw() +
    theme(strip.background = element_blank()) +
    ylab("Total passing touchdowns") +
    xlab("Week of season")
weekly_passing_id_r_plot
```

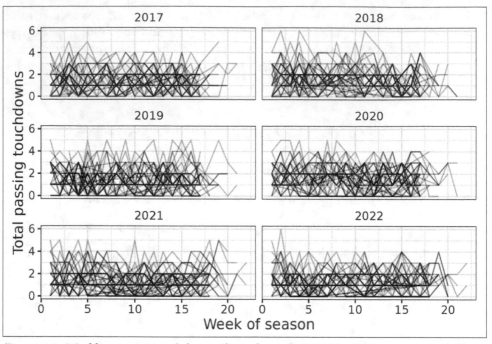

Figure 6-3. Weekly passing touchdowns throughout the 2017 to 2022 seasons. Each line corresponds to an individual passer.

Figure 6-3 shows the variability in the passing touchdowns per game. The values seem to be constant through time, and no trends appear to emerge. Add a Poisson regression trend line to the plot to create Figure 6-4:

```R
## R
weekly_passing_id_r_plot +
    geom_smooth(method = 'glm', method.args = list("family" = "poisson"),
                se=FALSE,
                linewidth = 0.5, color = 'blue',
                alpha = 0.25)
```

At first glance, no trends emerge in Figure 6-4. Players generally have a stable expected total passing touchdowns per game over the course of the seasons, with substantial variation week to week. Next, you'll investigate this data by using a model.

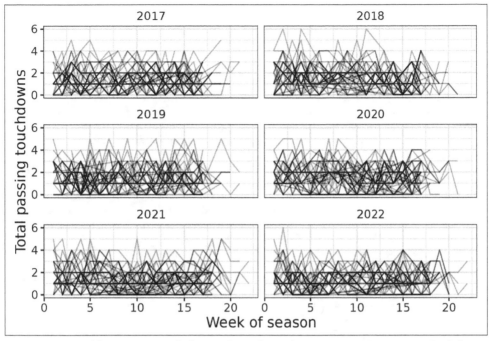

Figure 6-4. Weekly passing touchdowns throughout the 2017 to 2022 seasons. Each line corresponds to an individual passer. The trendline is from a Poisson regression.

Figures 6-3 and 6-4 do not provide much insight. We include them in this book to help you see the process of using exploratory data analysis (EDA) to check your data as you go. Most likely, these figures would not be used in communication unless a client specifically asked if a trend existed through time or you were writing a long, technical report or homework assignment.

Since you are assuming a Poisson distribution for the number of touchdown passes thrown in a game by a player, you use a *Poisson regression* as your model. The code to fit a Poisson regression is similar to a logistic regression from Chapter 5. However, the `family` is now Poisson rather than binomial. Poisson is required to do a Poisson regression. In Python, use this code to fit the model, save the outputs to the column in the data called `exp_pass_td`, and look at the summary:

```
## Python
pass_fit_py = \
    smf.glm(
        formula="pass_td_y ~ pass_td_rate + total_line",
        data=pbp_py_pass_td_y_geq10,
        family=sm.families.Poisson())\
    .fit()
```

```
pbp_py_pass_td_y_geq10["exp_pass_td"] = \
    pass_fit_py\
    .predict()

print(pass_fit_py.summary())
```

Resulting in:

```
               Generalized Linear Model Regression Results
==============================================================================
Dep. Variable:             pass_td_y   No. Observations:              3297
Model:                           GLM   Df Residuals:                  3294
Model Family:                Poisson   Df Model:                         2
Link Function:                   Log   Scale:                       1.0000
Method:                         IRLS   Log-Likelihood:              -4873.8
Date:                Sun, 04 Jun 2023   Deviance:                     3395.2
Time:                       09:41:29   Pearson chi2:                2.83e+03
No. Iterations:                    5   Pseudo R-squ. (CS):          0.07146
Covariance Type:           nonrobust
==============================================================================
                 coef    std err          z      P>|z|      [0.025      0.975]
------------------------------------------------------------------------------
Intercept      -0.9851      0.148     -6.641      0.000      -1.276      -0.694
pass_td_rate    0.3066      0.029     10.706      0.000       0.251       0.363
total_line      0.0196      0.003      5.660      0.000       0.013       0.026
==============================================================================
```

Likewise, in R use the following code to fit the model, save the outputs to the column in the data called exp_pass_td (note that you need to use type = "response" to put the output on the data scale rather than the coefficient/model scale), and look at the summary:

```
## R
pass_fit_r <-
    glm(pass_td_y ~ pass_td_rate + total_line,
        data = pbp_r_pass_td_y_geq10,
        family = "poisson"
    )

pbp_r_pass_td_y_geq10 <-
    pbp_r_pass_td_y_geq10 |>
    ungroup() |>
    mutate(exp_pass_td = predict(pass_fit_r, type = "response"))

summary(pass_fit_r) |>
    print()
```

Resulting in:

```
Call:
glm(formula = pass_td_y ~ pass_td_rate + total_line, family = "poisson",
    data = pbp_r_pass_td_y_geq10)

Coefficients:
             Estimate Std. Error z value Pr(>|z|)
(Intercept) -0.985076   0.148333  -6.641 3.12e-11 ***
pass_td_rate 0.306646   0.028643  10.706  < 2e-16 ***
total_line   0.019598   0.003463   5.660 1.52e-08 ***
---
Signif. codes:  0 '***' 0.001 '**' 0.01 '*' 0.05 '.' 0.1 ' ' 1

(Dispersion parameter for poisson family taken to be 1)

    Null deviance: 3639.6  on 3296  degrees of freedom
Residual deviance: 3395.2  on 3294  degrees of freedom
AIC: 9753.5

Number of Fisher Scoring iterations: 5
```

The coefficients and predictions from a Poisson regression depend on the scale for the mathematical link function, just like the logistic regression in "A Brief Primer on Odds Ratios" on page 132. "Poisson Regression Coefficients" on page 162 briefly explains the outputs from a Poisson regression.

Look at the coefficients and let's interpret them here. For the Poisson regression, coefficients are on an exponential scale (see the preceding warning on this topic for extra details). In Python, access the model's parameters and then take the exponential by using the NumPy library (np.exp()):

```
## Python
np.exp(pass_fit_py.params)
```

Resulting in:

```
Intercept       0.373411
pass_td_rate    1.358860
total_line      1.019791
dtype: float64
```

In R, use the tidy() function to look at the coefficients:

```
## R
library(broom)
tidy(pass_fit_r, exponentiate = TRUE, conf.int = TRUE)
```

Resulting in:

```
# A tibble: 3 × 7
  term          estimate std.error statistic  p.value conf.low conf.high
  <chr>            <dbl>     <dbl>     <dbl>    <dbl>    <dbl>     <dbl>
1 (Intercept)      0.373    0.148      -6.64 3.12e-11    0.279     0.499
2 pass_td_rate     1.36     0.0286     10.7  9.55e-27    1.28      1.44
3 total_line       1.02     0.00346     5.66 1.52e- 8    1.01      1.03
```

First, look at the `pass_td_rate` coefficient. For this coefficient, multiply every additional touchdown pass in the player's history by 1.36 to get the expected number of touchdown passes. Second, look at the `total_line` coefficient. For this coefficient, multiply the total line by 1.02. In this case, the total line is pretty efficient and is within 2% of the expected value (1 – 1.02 = –0.02). Both coefficients differ statistically from 0 on the (additive) model scale or differ statistically from 1 on the (multiplicative) data scale.

Now, look at Mahomes's data from Super Bowl LVII, leaving out the actual result for now (as well as the `passer_id` to save space) in Python:

```python
## Python
# specify filter criteria on own line for space
filter_by = 'passer == "P.Mahomes" & season == 2022 & week == 22'
# specify columns on own line for space
cols_look = [
    "season",
    "week",
    "passer",
    "total_line",
    "n_games",
    "pass_td_rate",
    "exp_pass_td",
]

pbp_py_pass_td_y_geq10.query(filter_by)[cols_look]
```

Resulting in:

```
      season  week    passer  total_line  n_games  pass_td_rate  exp_pass_td
3295    2022    22  P.Mahomes        51.0       39      2.384615     2.107833
```

Or look at the data in R:

```r
## R
pbp_r_pass_td_y_geq10 |>
    filter(passer == "P.Mahomes",
           season == 2022, week == 22) |>
    select(-pass_td_y, -n_passes, -passer_id, - week, -season, -n_games)
```

Resulting in:

```
# A tibble: 1 × 4
  passer    total_line pass_td_rate exp_pass_td
  <chr>          <dbl>        <dbl>       <dbl>
1 P.Mahomes         51         2.38        2.11
```

Now, what are these numbers?

- n_games shows that there are a total of 39 games (21 from the 2022 season and 18 from the previous season) in the Mahomes sample that we are considering.

- pass_td_rate is the current average number of touchdown passes per game by Mahomes in the sample that we are considering.

- exp_pass_td is the expected number of touchdown passes from the model for Mahomes in the Super Bowl.

And what do the numbers mean? These numbers show that Mahomes is expected to go under his previous average, even though his game total of 51 is relatively high. Likely, part of the concept of expected *touchdown passes* results from the concept of *regression toward the mean*. When people are above average, statistical models expect them to decrease to be closer to average (and the converse is true as well: below average players are expected to increase to be closer to the average). Because Mahomes is the best quarterback in the game at this point, the model predicts he is more likely to have a decrease rather than an increase. See Chapter 3 for more discussion about *regression toward the mean*.

Now, the average number doesn't really tell you that much. It's under 2.5, but the betting market already makes under 2.5 the favorite outcome. The question you're trying to ask is "It is too much or too little of a favorite?" To do this, you can use exp_pass_td as Mahomes's λ value and compute these probabilities explicitly by using the probability mass function and cumulative density function.

A *probability mass function* (PMF) gives you the probability of a discrete event occurring, assuming a statistical distribution. With our example, this is the probability of a passer completing a single number of touchdown passes per game, such as one touchdown pass or three touchdown passes. A *cumulative density function* (CDF) gives you the sum of the probability of multiple events occurring, assuming a statistical distribution. With our example, this would be the probability of completing X or fewer touchdown passes. For example, using $X = 2$, this would be the probability of completing zero, one, and two touchdown passes in a given week.

Python uses relatively straightforward names for PMFs and CDFs. Simply append the name to a distribution. In Python, use poisson.pmf() to give the probability of zero, one, or two touchdown passes being scored in a game by a quarterback.

Use 1 - `poisson.cdf()` to calculate the probability of more than two touchdown passes being scored in a game by a quarterback:

```Python
## Python
pbp_py_pass_td_y_geq10["p_0_td"] = \
    poisson.pmf(k=0,
                mu=pbp_py_pass_td_y_geq10["exp_pass_td"])

pbp_py_pass_td_y_geq10["p_1_td"] = \
    poisson.pmf(k=1,
                mu=pbp_py_pass_td_y_geq10["exp_pass_td"])

pbp_py_pass_td_y_geq10["p_2_td"] = \
    poisson.pmf(k=2,
                mu=pbp_py_pass_td_y_geq10["exp_pass_td"])

pbp_py_pass_td_y_geq10["p_g2_td"] = \
    1 - poisson.cdf(k=2,
                    mu=pbp_py_pass_td_y_geq10["exp_pass_td"])
```

From Python, look at the outputs for Mahomes going into the "big game" (or the Super Bowl, for readers who aren't familiar with football):

```Python
## Python
# specify filter criteria on own line for space
filter_by = 'passer == "P.Mahomes" & season == 2022 & week == 22'

# specify columns on own line for space
cols_look = [
    "passer",
    "total_line",
    "n_games",
    "pass_td_rate",
    "exp_pass_td",
    "p_0_td",
    "p_1_td",
    "p_2_td",
    "p_g2_td",
]

pbp_py_pass_td_y_geq10\
    .query(filter_by)[cols_look]
```

Resulting in:

```
          passer  total_line  n_games  ...    p_1_td    p_2_td   p_g2_td
3295   P.Mahomes        51.0       39  ...  0.256104  0.269912  0.352483

[1 rows x 9 columns]
```

R uses more confusing names for statistical distributions. The `dpois()` function gives the PMF, and the d comes from *density* because continuous distributions (such as the normal distribution) have density rather than mass. The `ppois()` function gives the CDF. In R, use the `dpois()` function to give the probability of zero, one, or two touchdown passes being scored in a game by a quarterback. Use the `ppois()` function to calculate the probability of more than two touchdown passes being scored in a game by a quarterback:

```R
## R
pbp_r_pass_td_y_geq10 <-
    pbp_r_pass_td_y_geq10 |>
    mutate(
        p_0_td = dpois(x = 0,
                       lambda = exp_pass_td),
        p_1_td = dpois(x = 1,
                       lambda = exp_pass_td),
        p_2_td = dpois(x = 2,
                       lambda = exp_pass_td),
        p_g2_td = ppois(q = 2,
                        lambda = exp_pass_td,
                        lower.tail = FALSE)
    )
```

Then look at the outputs for Mahomes going into the big game from R:

```R
## R
pbp_r_pass_td_y_geq10 |>
    filter(passer == "P.Mahomes", season == 2022, week == 22) |>
    select(-pass_td_y, -n_games, -n_passes,
           -passer_id, -week, -season)
```

Resulting in:

```
# A tibble: 1 × 8
  passer     total_line pass_td_rate exp_pass_td p_0_td p_1_td p_2_td p_g2_td
  <chr>           <dbl>        <dbl>       <dbl>  <dbl>  <dbl>  <dbl>   <dbl>
1 P.Mahomes          51         2.38        2.11  0.122  0.256  0.270   0.352
```

Notice in these examples we use the values for Mahomes's λ rather than hardcoding the value. We do this for several reasons. First, it allows us to see where the number comes from rather than being a *magical mystery number* with an unknown source. Second, it allows the number to self-update if the data were to change. This abstraction also allows our code to more readily generalize to other situations if we need or want to reuse the code. As you can see from this example, calculating one number can often require the same amount of code because of the beauty of vectorization in Python (with `numpy`) and R. In *vectorization*, a computer language applies a function on a vector (such as a column), rather than on a single value (such as a scalar).

OK, so you've estimated a probability of 35.2% that Mahomes would throw three or more touchdown passes, which is under the 40% you need to make a bet on the over.[1] The 64.8% that he throws two or fewer touchdown passes is slightly less than the 64.9% we need to place a bet on the under.

As an astute reader, you will notice that this is why most sports bettors don't win in the long term. At least at DraftKings Sportsbook, the math tells you not to make a bet or to find a different sportsbook altogether! FanDuel, one of the other major sportsbooks in the US, had a different index, 1.5 touchdowns, and offered –205 on the over (67.2% breakeven) and +164 (37.9%) on the under, which again offered no value, as we made the likelihood of one or fewer touchdown passes 37.8%, and two or more at 62.2%, with the under again just slightly under the break-even probability for a bet.

Luckily for those involved, Mahomes went over both 1.5 and 2.5 touchdown passes en route to the MVP. So the moral of the story is generally, "In all but a few cases, don't bet."

Poisson Regression Coefficients

The coefficients from a GLM depend on the link function, similar to the logistic regression covered in "A Brief Primer on Odds Ratios" on page 132. By default in Python and R, the Poisson regression requires that you apply the exponential function to be on the same scale as the data. However, this changes the coefficients on the link scale from being additive (like linear regression coefficients) to being multiplicative (like logistic regression coefficients).

To demonstrate this property, we will first use a simulation to help you better understand Poisson regression. First, you'll consider 10 *draws* (or samples) from a Poisson distribution with a mean of 1 and save this to be object x. Then look at the values for x by using `print()`, as well as the mean.

> For those of you following along in both languages, the Python and R examples usually produce different random values (unless, by chance, the values are the same). This is because both languages use slightly different random-number generators, and, even if the languages were somehow using the same identical random-number-generator function, the function call to generate the random numbers would be different.

---

1 All betting odds in this chapter came from Betstamp (*https://betstamp.app*).

In Python, use this code to generate the random numbers:

```Python
## Python
from scipy.stats import poisson

x = poisson.rvs(mu=1, size=10)
```

Then print the numbers:

```Python
## Python
print(x)
```

Resulting in:

```
[1 1 1 2 1 1 1 4 1 1]
```

And print their mean:

```Python
## Python
print(x.mean())
```

Resulting in:

```
1.4
```

In R, use this code to generate the numbers:

```R
## R
x <- rpois(n = 10, lambda = 1)
```

And then print the numbers:

```R
## R
print(x)
```

Resulting in:

```
[1] 2 0 6 1 1 1 3 0 1 1
```

And then look at their mean:

```R
## R
print(mean(x))
```

Resulting in:

```
[1] 1.6
```

Next, fit a GLM with a global intercept and look at the coefficient on the model scale and the exponential scale. In Python, use this code:

```Python
# Python
import statsmodels.formula.api as smf
import statsmodels.api as sm
import numpy as np
import pandas as pd

# create dataframe for glm
```

```
df_py = pd.DataFrame({"x": x})

# fit GLM
glm_out_py = \
    smf.glm(formula="x ~ 1", data=df_py, family=sm.families.Poisson()).fit()

# Look at output on model scale
print(glm_out_py.params)

# Look at output on exponential scale
```

Resulting in:

```
Intercept    0.336472
dtype: float64

print(np.exp(glm_out_py.params))
```

Resulting in:

```
Intercept    1.4
dtype: float64
```

In R, use this code:

```
## R
library(broom)

# fit GLM
glm_out_r <-
    glm(x ~ 1, family = "poisson")

# Look at output on model scale
print(tidy(glm_out_r))
```

Resulting in:

```
# A tibble: 1 × 5
  term        estimate std.error statistic p.value
  <chr>          <dbl>     <dbl>     <dbl>   <dbl>
1 (Intercept)    0.470     0.250      1.88  0.0601

# Look at output on exponential scale
print(tidy(glm_out_r, exponentiate = TRUE))
```

Resulting in:

```
# A tibble: 1 × 5
  term        estimate std.error statistic p.value
  <chr>          <dbl>     <dbl>     <dbl>   <dbl>
1 (Intercept)     1.60     0.250      1.88  0.0601
```

Notice that the coefficient on the exponential scale (the second printed table, with exponentiate = TRUE) is the same as the mean of the simulated data. However, what if you have two coefficients, such as a slope and an intercept?

As a concrete example, consider the number of touchdowns per game for the Baltimore Ravens. During this season, the Ravens' quarterback was unfortunately injured during game 13. Hence, you might reasonably expect the Ravens' number of passes to decrease over the course of the season. A formal method to test this would be to ask the question "Does the average (or expected) number of touchdowns per game change through the season?" and then use a Poisson regression to statistically evaluate this.

To test this, first wrangle the data and then plot it to help you see what is going on with the data. You will also shift the week by subtracting 1. This allows week 1 to be the intercept of the model. You will also set the axis ticks and change the axis labels. In Python, use this code to create Figure 6-5:

```Python
## Python
# subset the data
bal_td_py = (
    pbp_py
    .query('posteam=="BAL" & season == 2022')
    .groupby(["game_id", "week"])
    .agg({"touchdown": ["sum"]})
)

# reformat the columns
bal_td_py.columns = list(map("_".join, bal_td_py.columns))
bal_td_py.reset_index(inplace=True)

# shift week so intercept 0 = week 1
bal_td_py["week"] = bal_td_py["week"] - 1

# create list of weeks for plot
weeks_plot = np.linspace(start=0, stop=18, num=10)
weeks_plot

# plot the data
```

Resulting in:

```
array([ 0.,  2.,  4.,  6.,  8., 10., 12., 14., 16., 18.])
```

```
ax = sns.regplot(data=bal_td_py, x="week", y="touchdown_sum");
ax.set_xticks(ticks = weeks_plot, labels = weeks_plot);
plt.xlabel("Week")
plt.ylabel("Touchdowns per game")

plt.show();
```

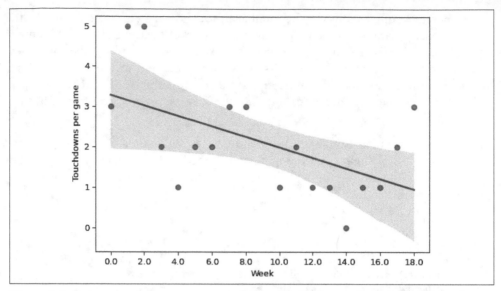

Figure 6-5. Ravens touchdowns per game during the 2022 season, plotted with seaborn. *Notice the use of a linear regression trendline.*

 Most Python plotting tools are wrappers for the matplotlib package, and seaborn is no exception. Hence, customizing seaborn will usually require the use of matplotlib commands, and if you want to become an expert in plotting with Python, you will likely need to become comfortable with the clunky matplotlib commands.

In R, use this code to create Figure 6-6:

```r
## r
bal_td_r <-
    pbp_r |>
    filter(posteam == "BAL" & season == 2022) |>
    group_by(game_id, week) |>
    summarize(
        td_per_game =
            sum(touchdown, na.rm = TRUE),
        .groups = "drop"
    ) |>
    mutate(week = week - 1)

ggplot(bal_td_r, aes(x = week, y = td_per_game)) +
    geom_point() +
    theme_bw() +
    stat_smooth(
        method = "glm", formula = "y ~ x",
        method.args = list(family = "poisson")
    ) +
```

```
xlab("Week") +
ylab("Touchdowns per game") +
scale_y_continuous(breaks = seq(0, 6)) +
scale_x_continuous(breaks = seq(1, 20, by = 2))
```

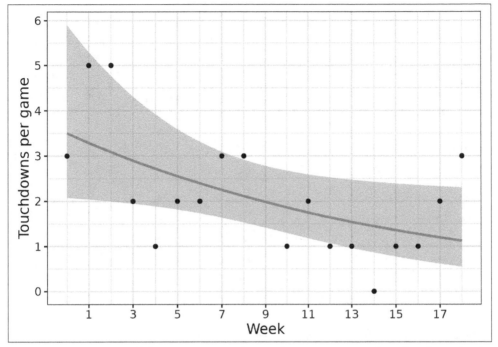

Figure 6-6. Ravens touchdowns per game during the 2022 season, plotted with ggplot2.
Notice the use of a Poisson regression trend line.

When comparing Figures 6-5 and 6-6, notice that ggplot2 allows you to use a
Poisson regression for the trendline, whereas seaborn allows only a liner model. Both
figures show that Baltimore, on average, has a decreasing number of touchdowns per
game. Now, let's build a model and look at the coefficients. Here's the code in Python:

```Python
## Python
glm_bal_td_py = \
    smf.glm(formula="touchdown_sum ~ week",
            data=bal_td_py,
            family=sm.families.Poisson())\
    .fit()
```

Then look at the coefficients on the link (or log) scale:

```
## Python
print(glm_bal_td_py.params)
```

Resulting in:

```
Intercept    1.253350
week        -0.063162
dtype: float64
```

And look at the exponential (or data) scale:

```
## Python
print(np.exp(glm_bal_td_py.params))
```

Resulting in:

```
Intercept    3.502055
week         0.938791
dtype: float64
```

Or use R to fit the model:

```
## R
glm_bal_td_r <-
    glm(td_per_game ~ week,
        data = bal_td_r,
        family = "poisson"
    )
```

Then look at the coefficient on the link (or log) scale:

```
## R
print(tidy(glm_bal_td_r))
```

Resulting in:

```
# A tibble: 2 × 5
  term        estimate std.error statistic   p.value
  <chr>          <dbl>     <dbl>     <dbl>     <dbl>
1 (Intercept)    1.25      0.267      4.70 0.00000260
2 week          -0.0632    0.0300    -2.10 0.0353
```

And look at the exponential (or data) scale:

```
## R
print(tidy(glm_bal_td_r, exponentiate = TRUE))
```

Resulting in:

```
# A tibble: 2 × 5
  term        estimate std.error statistic   p.value
  <chr>          <dbl>     <dbl>     <dbl>     <dbl>
1 (Intercept)    3.50      0.267      4.70 0.00000260
2 week           0.939     0.0300    -2.10 0.0353
```

Now, let's look the meaning of the coefficients. First, examining the link-scale value of the intercept, `1.253`, and the slope, `-0.063`, does not appear to provide much context. However, the slope term is also known as the *risk ratio*, or *relative risk*, for each week. After exponentiating the coefficients, you get an intercept value of 3.502 and a slope value of 0.939.

These numbers do have easy-to-understand values. The intercept is the number of expected passes during the first game (this is why we had you subtract 1 from week, so that week 1 would be 0 and the intercept), and its value is 3.502. Looking at Figures 6-5 and 6-6, this seems reasonable because week 0 was 3 and weeks 1 and 2 were 5.

 Using multiplication on the log scale is the same as using addition on nontransformed numbers. For example, $x + y = \log(e^x \times e^y)$. To demonstrate with real numbers, consider $1 + 3 = 4$. And (ignoring rounding errors), $e^1 = 2.718$ and $e^3 = 20.09$. Using these results, $2.718 \times 20.09 = 54.60$. Lastly, back transforming gives you: $\log(54.60) = 4.00$.

Then, for the next week, the expected number of touchdowns would be $3.502 \times 0.939 = 3.288$. For the following week, $3.502 \times 0.939^2 = 3.088$. Hence, the equation to estimate the expected number of touchdowns for a given week during the 2022 season for the Baltimore Ravens would be $3.502 \times 0.939^{\text{week}}$ on the log scale. Or this could be rewritten as $e^{1.253 + -0.063 \times \text{week}}$.

Although we had to cherry-pick this example to explain a Poisson regression, hopefully the example served its purpose and helps you see how to interpret Poisson regression terms.

Closing Thoughts on GLMs

This chapter and Chapter 5 have introduced you to GLMs and how the models can help you both understand and predict with data. In fact, the simple predictions done in these chapters would help you also understand the basics of simple machine learning methods. When working with GLMs, notice that both example models forced you to deal with the link function because the coefficients are estimated on a different scale compared to the data. Additionally, you saw two types of error families, the binomial in Chapter 5 and the Poisson in this chapter.

In the exercises in this chapter, we have you look at the quasi-Poisson to account for dispersion when there are either too many or too few 0s compared to what would be expected. We also have you look at the negative binomial as another alternative to Poisson for modeling count data. Two other common types of GLMs include the gamma regression and the lognormal regression. The gamma and lognormal are similar. The lognormal might be better for some data that is positive but has a *right skew* as pass data (the data has more or greater positive values on the right side of the distribution and has smaller or negative values on the left side of the distribution).

An alphabet soup of extensions exists for linear models and GLMS. Hierarchical, multilevel, or random-effect models allow for features such as repeated observations on the same individual or group to model uncertainty. Generalized additive models (GAMs) allow for curves, rather than straight lines, to be fit for models. Generalized additive mixed-effect models (GAMMs) merge GAMs and random-effect models. Likewise, tools such as model selection can help you determine which type of model to use.

Basically, linear models have many options because of their long history and use in statistics and scientific fields. We have a colleague who can, and would, talk about the assumptions of regressions for hours on end. Likewise, year-long graduate-level courses exist on the models we have covered over the course of a few chapters. However, even simply knowing the basics of regression models will greatly help you to up your football analytics game.

Data Science Tools Used in This Chapter

This chapter covered the following topics:

- Fitting a Poisson regression in Python and R by using `glm()`
- Understanding and reading the coefficients from a Poisson regression including relative risk
- Connecting to datasets by using `merge()` in Python or an `inner_join()` in R
- Reapplying data-wrangling tools you learned in previous chapters
- Learning about betting odds

Exercises

1. What happens in the model for touchdown passes if you don't include the total for the game? Does it change any of the probabilities enough to recommend a bet for Patrick Mahomes's touchdown passes in the Super Bowl?

2. Repeat the work in this chapter for interceptions thrown. Examine Mahomes's interception prop in Super Bowl LVII, which was 0.5 interceptions, with the over priced at –120 and the under priced at –110.

3. What about Jalen Hurts, the other quarterback in Super Bowl LVII, whose touchdown prop was 1.5 (–115 to the over and –115 to the under) and interception prop was 0.5 (+105/–135)? Was there value in betting either of those markets?

4. Look through the `nflfastR` dataset for additional features to add to the models for both touchdowns and interceptions. Does pregame weather affect either total? Does the size of the point spread?

5. Repeat this chapter's GLMS, but rather than using a Poisson distribution, use a quasi-Poisson that estimates dispersion parameters.

6. Repeat this chapter's GLMS, but rather than using Poisson distribution, use a negative binomial.

Suggested Readings

The resources suggested in "Suggested Readings" on pages 76, 112, and 136 will help you understand generalized linear models such as Poisson regression more. One last regression book you may find helpful is: *Applied Generalized Linear Models and Multilevel Models in R* by Paul Roback and Julie Legler (CRC Press, 2021). This book provides an accessible introduction for people wanting to learn more about regression analysis and advanced tools such as GLMs.

To learn more about the application of probability to both betting and the broader world, check out the following resources:

- *The Logic of Sports Betting* by Ed Miller and Matthew Davidow (self-published, 2019) provides details about how betting works for those wanting more details.

- *The Wisdom of the Crowds* by James Surowiecki (Doubleday, 2004) describes how "the market" or collective guess of the crowds can predict outcomes.

- *Sharp Sports Betting* by Stanford Wong (Huntington Press, 2021) will help readers understand betting more and provides an introduction to the topic.

- *Sharper: a Guide to Modern Sports Betting* by True Pokerjo (self-published, 2016) talks about sports betting for those seeking insight.

- *The Foundations of Statistics,* 2nd revised edition, by Leonard J. Savage (Dover Press, 1972) is a classical statistical text on how to use statistics to make decisions. The first half of the book does a good job explaining statistics in real-world application. The second becomes theoretical for those seeking a mathematically rigorous foundation. (Richard skimmed and then skipped much of the second half.)

- The Good Judgment Project (*https://goodjudgment.com*) seeks to apply the "wisdom of the crowds" to predict real-world problems. Initially funded by the US Intelligence Community, this project's website states that its "forecasts were so accurate that they even outperformed intelligence analysts with access to classified data."

## CHAPTER 7

Web Scraping: Obtaining and Analyzing Draft Picks

One of the great triumphs in public analysis of American football is nflscrapR and, after that, nflfastR. These packages allow for easy analysis of the game we all love. Including data in your computing space is often as simple as downloading a package in Python or R, and away you go.

Sometimes it's not that easy, though. Often you need to *scrape* data off the web yourself (use a computer program to download your data). While it is beyond the scope of this book to teach you all of web scraping in Python and R, some pretty easy commands can get you a significant amount of data to analyze.

In this chapter, you are going to scrape NFL Draft and NFL Scouting Combine data from Pro Football Reference (*https://www.pro-football-reference.com*). It's a wonderful resource out of Philadelphia, Pennsylvania. It's owned by Sports Reference, which also provides free data for every sport imaginable. You will use this website to get data for the NFL Draft and NFL Scouting Combine.

The *NFL Draft* is a yearly event held in various cities around the country. In the draft, teams select from a pool of players who have completed at least three post–high school years. While it used to have more rounds, the NFL Draft currently consists of seven rounds. The draft order in each round is determined by how well each team played the year before. Weaker teams pick higher in the draft than the stronger teams. Teams can trade draft picks for other draft picks or players.

The *NFL Scouting Combine* is a yearly event held in Indianapolis, Indiana. In the combine, a pool of athletes eligible for the NFL Draft meet with evaluators from NFL teams to perform various physical and psychological tests. Additionally, this is generally thought of as the NFL's yearly convention, where deals between teams and agents are originated and, sometimes, finalized.

The combination of these two datasets is a great resource for beginners in football analytics for a couple of reasons. First, the data is collected over a few days once a year and does not change thereafter. Although some players may retest physically at a later date, and players can often leave the team that drafted them for various reasons, the draft teams cannot change. Thus, once you obtain the data, it's generally good to use for almost an entire calendar year, after which you can simply add the new data when it's obtained the following year.

You will start by scraping all NFL Scouting Combine and NFL Draft data from 2022 and then fold in later years for analysis.

 Web scraping involves a lot of trial and error, especially when you're getting started. In general, we find an example that works and then change one piece at a time until we get what we need.

Web Scraping with Python

 Before you start web scraping, go to the web page first so you can see what you are trying to download.

The following code allows us to scrape with Python by using `for` loops. If you have skipped chapters or require a reminder, "Individual Player Markets and Modeling" on page 149 provides an introduction to `for` loops. Save the *uniform resource locator (URL)* or web address, to an object, `url`. In this case, the URL is simply the URL for the 2022 NFL Draft.

Next, use `read_html()` from the `pandas` package to simply read in tables from the given URL. Remember that Python starts counting with 0. Thus, the zeroth element of the dataframe, `draft_py` from `read_html()`, is simply the first table on the web page. You will also need to change `NA` draft approximate values to be `0`:

```
## Python
import pandas as pd
import seaborn as sns
import matplotlib.pyplot as plt
import statsmodels.formula.api as smf
import numpy as np

url = "https://www.pro-football-reference.com/years/2022/draft.htm"

draft_py = pd.read_html(url, header=1)[0]
draft_py.loc[draft_py["DrAV"].isnull(), "DrAV"] = 0
```

You can peek at the data by using `print()`:

```
## Python
print(draft_py)
```

Resulting in:

```
     Rnd Pick   Tm   Sk  College/Univ   Unnamed: 28
0      1    1  JAX  3.5       Georgia  College Stats
1      1    2  DET  9.5      Michigan  College Stats
2      1    3  HOU  1.0           LSU  College Stats
3      1    4  NYJ  NaN    Cincinnati  College Stats
4      1    5  NYG  4.0        Oregon  College Stats
..    ..  ...  ...  ...           ...            ...
263    7  258  GNB  NaN      Nebraska  College Stats
264    7  259  KAN  NaN      Marshall  College Stats
265    7  260  LAC  NaN        Purdue  College Stats
266    7  261  LAR  NaN  Michigan St.  College Stats
267    7  262  SFO  NaN      Iowa St.  College Stats

[5871 rows x 31 columns]
```

 When web scraping, be careful not to *hit*, or pull from, websites too many times. You may find yourself locked out of websites. If this occurs, you will need to wait a while until you try again. Additionally, many websites have rules (more formally known as *Terms & Conditions*) that provide guidance on whether and how you can scrape their pages.

Although kind of ugly, this web-scraping process is workable! To scrape multiple years (for example, 2000 to 2022), you can use a simple `for` loop—which is often possible because of systematic changes in the data. Experimentation is key.

To avoid multiple pulls from web pages, we cached files when writing the book and downloaded them only when needed. For example, we used (and hid from you, the reader) this code earlier in this chapter:

```python
## Python
import pandas as pd
import os.path

file_name = "draft_demo_py.csv"

if not os.path.isfile(file_name):
    ## Python
    url = \
        "https://www.pro-football-reference.com/" + \
        "years/2022/draft.htm"
    draft_py = pd.read_html(url, header=1)[0]

    conditions = [
        (draft_py.Tm == "SDG"),
        (draft_py.Tm == "OAK"),
        (draft_py.Tm == "STL"),
    ]
    choices = ["LAC", "LV", "LAR"]

    draft_py["Tm"] = \
        np.select(conditions, choices,
                    default = draft_py.Tm)
    draft_py.loc[draft_py["DrAV"].isnull(), "DrAV"] = 0
    draft_py.to_csv(file_name)
else:
    draft_py = pd.read_csv(file_name)
    draft_py.loc[draft_py["DrAV"].isnull(), "DrAV"] = 0
```

Also, when creating our own for loops, we often start with a simple index value (for example, set i = 1) and then make our code work. After making our code work, we add in the for line to run the code over many values.

When setting the index value to a value such as 1 while building for loops, make sure you remove the placeholder index (such as i = 1) from your code. Otherwise, your loop will simply run over the same functions or data multiple times. We have made this mistake when coding more times than we would like to admit.

Now, let's download more data in Python. As part of this process, you need to clean up the data. This includes telling pandas which row has the header—in this case, the second row, or header=1. Recall that Python starts counting with 0, so 1 corresponds to the second entry. Likewise, you need to save the season as part of the loop to its own columns. Also, remove rows that contain extra heading information (strangely, some rows in the dataset are duplicates of the data's header) by saving only rows whose value is not equal to the column's name (for example, use tm != "Tm"):

```
## Python
draft_py = pd.DataFrame()
for i in range(2000, 2022 + 1):
    url = "https://www.pro-football-reference.com/years/" + \
            str(i) + \
            "/draft.htm"
    web_data = pd.read_html(url, header=1)[0]
    web_data["Season"] = i
    web_data = web_data.query('Tm != "Tm"')
    draft_py = pd.concat([draft_py, web_data])

draft_py.reset_index(drop=True, inplace=True)
```

Some teams moved cities over the past decade; therefore, the team names (Tm) need to be changed to reflect the new locations. The np.select() function can be used for this, using conditions that has the old names, and choices that has the new names. The default in np.select() also needs to be changed so that teams that haven't moved stay the same:

```
## Python
# the Chargers moved to Los Angeles from San Diego
# the Raiders moved from Oakland to Las Vegas
# the Rams moved from St. Louis to Los Angeles
conditions = [
    (draft_py.Tm == "SDG"),
    (draft_py.Tm == "OAK"),
    (draft_py.Tm == "STL"),
]
choices = ["LAC", "LVR", "LAR"]

draft_py["Tm"] = \
    np.select(conditions, choices, default = draft_py.Tm)
```

Finally, replace missing draft approximate values with 0 before you reset the index and save the file:

```
## Python
draft_py.loc[draft_py["DrAV"].isnull(), "DrAV"] = 0
draft_py.to_csv("data_py.csv", index=False)
```

Now, you can peek at the data:

```
## Python
print(draft_py.head())
```

Resulting in:

```
   Unnamed: 0  Rnd  Pick  Tm  ...    Sk College/Univ    Unnamed: 28  Season
0           0    1     1  CLE ...  19.0    Penn St.  College Stats     2000
1           1    1     2  WAS ...  23.5    Penn St.  College Stats     2000
2           2    1     3  WAS ...   NaN     Alabama  College Stats     2000
3           3    1     4  CIN ...   NaN  Florida St.  College Stats     2000
4           4    1     5  BAL ...   NaN   Tennessee  College Stats     2000

[5 rows x 31 columns]
```

Let's look at the other columns available to us:

```
## Python
print(draft_py.columns)
```

Resulting in:

```
Index(['Unnamed: 0', 'Rnd', 'Pick', 'Tm', 'Player', 'Pos', 'Age', 'To', 'AP1',
       'PB', 'St', 'wAV', 'DrAV', 'G', 'Cmp', 'Att', 'Yds', 'TD', 'Int',
       'Att.1', 'Yds.1', 'TD.1', 'Rec', 'Yds.2', 'TD.2', 'Solo', 'Int.1', 'Sk',
       'College/Univ', 'Unnamed: 28', 'Season'],
      dtype='object')
```

With R and Python, we usually need to tell the computer to save our updates. Hence, we often save objects over the same name, such as draft_py = draft_py.drop(labels = 0, axis = 0). In general, understand this copy behavior so that you do not delete data you want or need later.

Finally, here's the metadata, or *data dictionary* (data about data) for the data you will care about for the purposes of this analysis:

- The season in which the player was drafted (Season)
- Which selection number they were taken at (Pick)
- The player's drafting team (Tm)
- The player's name (Player)
- The player's position (Pos)
- The player's whole career approximate value (wAv)
- The player's approximate value for the drafting team (DrAV)

Lastly, you might want to reorder and select on certain columns. For example, you might want only six columns and to change their order:

```
## Python
draft_py_use = \
    draft_py[["Season", "Pick", "Tm", "Player", "Pos", "wAV", "DrAV"]]

print(draft_py_use)
```

Resulting in:

```
      Season  Pick   Tm           Player  Pos   wAV  DrAV
0       2000     1  CLE    Courtney Brown   DE  27.0  21.0
1       2000     2  WAS  LaVar Arrington   LB  46.0  45.0
2       2000     3  WAS    Chris Samuels    T  63.0  63.0
3       2000     4  CIN    Peter Warrick   WR  27.0  25.0
4       2000     5  BAL      Jamal Lewis   RB  69.0  53.0
...      ...   ...  ...              ...  ...   ...   ...
5866    2022   258  GNB     Samori Toure   WR   1.0   1.0
5867    2022   259  KAN   Nazeeh Johnson  SAF   1.0   1.0
5868    2022   260  LAC   Zander Horvath   RB   0.0   0.0
5869    2022   261  LAR        AJ Arcuri   OT   1.0   1.0
5870    2022   262  SFO      Brock Purdy   QB   6.0   6.0

[5871 rows x 7 columns]
```

You'll generally want to save this data locally for future use. You do not want to download and clean data each time you use it.

The data you obtain from web scraping may be different from our example data. For example, a technical reviewer had the *draft pick* column treated as a discrete character rather than a continuous integer or numeric. If your data seems strange (such as plots that seem off), examine your data by using tools from Chapter 2 and other chapters to check your data types.

Web Scraping in R

Some Python and R packages require outside dependencies, especially on macOS and Linux. If you try to install a package and get an error message, try reading the error message. Often we find the error messages to be cryptic, so we end up using a search engine to help our debugging.

You can use the `rvest` package to create similar loops in R. First, load the package and create an empty `tibble`:

```
## R
library(janitor)
library(tidyverse)
library(rvest)
library(htmlTable)
library(zoo)

draft_r <- tibble()
```

Then loop over the years 2000 to 2022. Ranges can be specified using a colon, such as `2000:2022`. However, we prefer to explicitly use the `seq()` command because it is more robust.

A key difference in the R code compared to the Python code is that the `html_nodes` command is called with the pipe. The code also needs to extract the web dataframe (`web_df`) from the raw `web_data`. The `row_to_names()` function cleans up for the empty header row and replaces the data's header with the first row. The `janitor::clean_names()` function cleans up the column names because some columns have duplicate names. The `mutate()` function saves the season in the loop.

Next, use `filter()` with the data to remove any rows that contain duplicate headers as extra rows:

```
## R

for (i in seq(from = 2000, to = 2022)) {
    url <- paste0(
        "https://www.pro-football-reference.com/years/",
        i,
        "/draft.htm"
    )
    web_data <-
        read_html(url) |>
        html_nodes(xpath = '//*[@id="drafts"]') |>
        html_table()
    web_df <-
        web_data[[1]]
    web_df_clean <-
        web_df |>
        janitor::row_to_names(row_number = 1) |>
        janitor::clean_names(case = "none") |>
        mutate(Season = i) |> # add seasons
        filter(Tm != "Tm") # Remove any extra column headers

    draft_r <-
        bind_rows(
            draft_r,
            web_df_clean
        )
}
```

Rename teams (Tm) to reflect those that have moved by using `case_when()` before saving the output; now you do not need to download it again. You also save and reload the data because this further cleans the data for R.

```
## R
# the chargers moved to Los Angeles from San Diego
# the Raiders moved from Oakland to Las Vegas
# the Rams moved from St. Louis to Los Angeles
draft_r <-
    draft_r |>
    mutate(Tm = case_when(Tm == "SDG" ~ "LAC",
                          Tm == "OAK" ~ "LVR",
                          Tm == "STL" ~ "LAR",
                          TRUE ~ Tm),
           DrAV = ifelse(is.na(DrAV), 0, DrAV))
write_csv(draft_r, "draft_data_r.csv")
draft_r <- read_csv( "draft_data_r.csv")
```

Now that you have data, use **select()** to grab the data you'll need for the analysis later:

```
## R
draft_r_use <-
    draft_r |>
    select(Season, Pick, Tm, Player, Pos, wAV, DrAV)

print(draft_r_use)
```

Resulting in:

```
# A tibble: 5,871 × 7
   Season  Pick Tm    Player            Pos    wAV  DrAV
    <dbl> <dbl> <chr> <chr>             <chr> <dbl> <dbl>
 1   2000     1 CLE   Courtney Brown    DE       27    21
 2   2000     2 WAS   LaVar Arrington   LB       46    45
 3   2000     3 WAS   Chris Samuels     T        63    63
 4   2000     4 CIN   Peter Warrick     WR       27    25
 5   2000     5 BAL   Jamal Lewis       RB       69    53
 6   2000     6 PHI   Corey Simon       DT       45    41
 7   2000     7 ARI   Thomas Jones      RB       62     7
 8   2000     8 PIT   Plaxico Burress   WR       70    34
 9   2000     9 CHI   Brian Urlacher HOF LB     119   119
10   2000    10 BAL   Travis Taylor     WR       30    23
# i 5,861 more rows
```

Compare the two web-scraping methods. Python functions tend to be more self-contained and tend to belong to an object (and functions that belong to an object in Python are called *methods*). In contrast, R tends to use multiple functions on the same object. This is a design trait of the languages. Python is a more object-oriented language, whereas R is a more functional language. Which style do you like better?

You might notice one thing here that won't affect the analysis in this chapter, but will if you want to move forward with the data: the ninth pick in the 2000 NFL Draft has the name *Brian Urlacher HOF*. The *HOF* part denotes that he eventually made the NFL Hall of Fame. If you want to use this data and merge it with another dataset, you will have to alter the names to make sure that details like that are taken out.

Analyzing the NFL Draft

The NFL Draft occurs annually and allows teams to select eligible players. The event started in 1936 as a way to allow all teams to remain competitive and obtain talented players. During the draft, each team gets one pick per round. The order for each round is based on a team's record, with tie-breaking rules for teams with the same record. Thus, the team that won the Super Bowl picks last, and the team that lost the Super Bowl picks second-to-last. However, teams often trade players and include draft picks as part of their trades. Hence picks can have extra value that people want to quantify and understand.

You can ask many cool questions about the draft, especially if you're drafting players (either for a fantasy team or a real team). How much is each draft pick worth (and in what denomination)? Are some teams better at drafting players than others? Are some positions better gambles than others in the draft?

At first blush, the most straightforward question to answer with this data is the first question—"How much is each draft pick worth (and in what denomination)?"—which is assigning a value to each draft pick. The reason this is important is that teams will often trade picks to each other in an effort to align the utility of their draft picks with the team's current needs.

For example, the New York Jets, which was holding the sixth-overall pick in the 2018 draft, traded that pick, along with the 37th and 49th picks in the 2018 draft, as well as the second-round pick in the 2019 draft, to the Indianapolis Colts for the third pick in 2018. Top draft picks are generally reserved for quarterbacks, and the Jets, after losing out on the Kirk Cousins sweepstakes in free agency, needed a quarterback. The Colts "earned" the third-overall pick in 2018 in large part because the team struggled in 2017 with its franchise quarterback, Andrew Luck, who was on the mend with a shoulder injury. In other words the third-overall pick had less utility to the Colts than it held for the Jets, so the teams made the trade.

How do the teams decide what is a "fair" market value for these picks? You have to go back to 1989, when the Dallas Cowboys, after being bought by Arkansas oil millionaire Jerry Jones, jettisoned long-time Hall of Fame head coach Tom Landry and replaced him with a college coach, Jimmy Johnson. The Cowboys were coming off a three-win season in 1988, and their roster was relatively bare when it came to difference-making players. As the legend goes, Johnson, while on a jog, decided to trade his star running back (and future US senatorial candidate) Herschel Walker for a package that ended up being three first-round picks, three-second round picks, a third-round pick, a sixth-round pick, and a few players.

The Cowboys would finish 1989 with the NFL's worst record at 1–15, which improved the position of their remaining, natural draft picks (the team used its 1990 first-rounder early on a quarterback in the supplemental draft). Johnson would go on to make more draft-pick trades during the early years of his tenure than any other coach or executive in the game, using a value chart made by Mike McCoy, which is now known as the *Jimmy Johnson chart*, designed to match the value of each pick in each round. The chart assigns the first-overall pick 3,000 points, with the value of each subsequent pick falling off exponentially.

The Jimmy Johnson chart is still the chart of choice for many NFL teams but has been shown to overvalue top picks in the draft relative to subsequent picks. In fact, when considering the cost of signing each pick, Nobel Prize–winning economist and sports analytics legend Cade Massey showed in "The Loser's Curse: Decision Making and Market Efficiency in the National Football League Draft" (*https://oreil.ly/vFGt6*) that the surplus value of the first pick, that is, the on-field value of the pick relative to the salary earned by the player picked, is not maximized by the first pick but rather by picks either in the middle or end of the first-round picks or early in the second-round picks.

Much commentary exists on this topic, some of it useful, and some of it interactive via code. Eric's former colleague at PFF, Timo Riske, wrote an article (*https://oreil.ly/7J9HO*) about "the surplus value of draft picks." Additionally, others have done research on this topic, including Michael Lopez (*https://oreil.ly/e0mMH*) and Ben Baldwin (*https://oreil.ly/eR4Ra*), the same Ben Baldwin who helped create nflfastR, and have produced even shallower draft curves.

Here we aim to have you reproduce that research by using the amount of approximate value generated by each player picked for his drafting team, or *draft approximate value (DrAV)*. Plotting this, you can clearly see that teams drafting future players have market efficiency, because teams get better picks earlier in the draft (some bias exists here, in that teams also play their high draft picks more, but you can show that per-play efficiency drops off with draft slot as well).

First, select years prior to 2019. The reason to filter out years after 2019 is that those players are still playing through their rookie contracts and haven't yet had a chance to sign freely with their next team.

Then plot the data to compare each pick to its average draft value. In Python, use this code to create Figure 7-1:

```python
## Python
# Change theme for chapter
sns.set_theme(style="whitegrid", palette="colorblind")

draft_py_use_pre2019 = \
    draft_py_use\
    .query("Season <= 2019")

## format columns as numeric or integers
draft_py_use_pre2019 = \
    draft_py_use_pre2019\
    .astype({"Pick": int, "DrAV": float})

sns.regplot(data=draft_py_use_pre2019,
            x="Pick",
            y="DrAV",
            line_kws={"color": "red"},
            scatter_kws={'alpha':0.2});
plt.show();
```

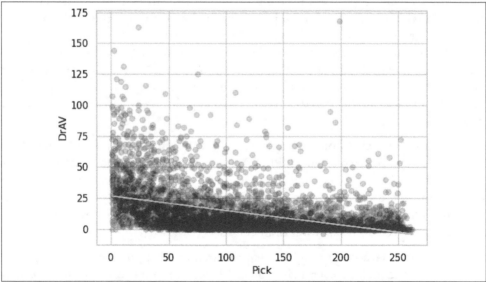

Figure 7-1. Scatterplot with a linear trendline for draft pick number against draft approximate value, plotted with seaborn

In R, use this code to create Figure 7-2:

```
## R
draft_r_use_pre2019 <-
    draft_r_use |>
    mutate(DrAV = as.numeric(DrAV),
           wAV = as.numeric(wAV),
           Pick = as.integer(Pick)) |>
    filter(Season <= 2019)

ggplot(draft_r_use_pre2019, aes(Pick, DrAV)) +
    geom_point(alpha = 0.2) +
    stat_smooth() +
    theme_bw()
```

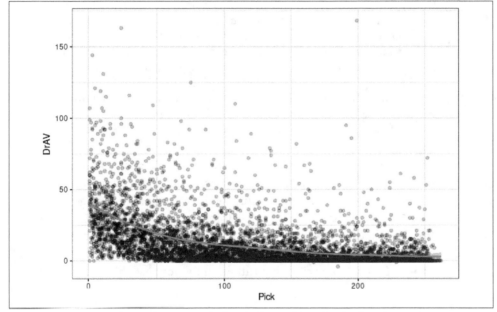

Figure 7-2. Scatterplot with a smoothed, spline trendline for draft pick number against draft approximate value, plotted with ggplot2

Figures 7-1 and 7-2 show that the value of a pick decreases as the pick number increases.

Now, a real question when trying to derive this curve is what are teams looking for when they draft? Are they looking for the average value produced by the pick? Are they looking for the median value produced by the pick? Some other percentile?

The median is likely going to give 0 values for later picks, which is clearly not true since teams trade them all the time, but the mean might also overvalue them because of the hits later in the draft by some teams. Statistically, the median is 0 if more than

50% of picks had 0 value. But the mean can be affected by some really good players who were drafted late. The Patriots, for example, took Tom Brady with the 199th pick in the 2000 draft, and he became one of the best football players of all time. A *Business Insider* article (*https://oreil.ly/wqcHe*) by Cork Gaines tells the backstory of this pick.

For now, use the mean, but in the exercises you'll use the median and see if anything changes. "Quantile Regression" on page 242 provides a brief overview of another type of regression, quantile regression, that might be worth exploring if you want to dive into different model types.

 Datasets that contain series (such as daily temperatures or draft picks) often contain both patterns and noise. One way to smooth out this noise is by calculating an average over several sequential observations. This average is often called a *rolling average*; other names include *moving mean*, *running average*, or similar variations with *rolling*, *moving*, or *running* used to describe the mean or average. Key inputs to a rolling average include the *window* (number of inputs to use), method (such as mean or median), and what to do with the start and end of the series (for example, should the first entries without a full window be dropped or another rule used?).

To smooth out the value for each pick, first calculate the average value for each pick. A couple of the lower picks had NaN values, so replace these values with 0. Then, calculate the six-pick moving mean DrAV surrounding each pick's average value (that is to say, each DrAV value as well as the 6 before and 6 after, for a window of 13). Also, tell the rolling() function to use min_periods=1 and to center the mean (the rolling average is centered on the current DrAV). Last, groupby() pick and then calculate the average DrAV for each pick position. In Python, use this code:

```
## Python
draft_chart_py = \
    draft_py_use_pre2019\
    .groupby(["Pick"])\
    .agg({"DrAV": ["mean"]})

draft_chart_py.columns = \
    list(map("_".join, draft_chart_py.columns))

draft_chart_py.loc[draft_chart_py.DrAV_mean.isnull()] = 0

draft_chart_py["roll_DrAV"] = (
    draft_chart_py["DrAV_mean"]
    .rolling(window=13, min_periods=1, center=True)
    .mean()
)
```

For the exercise in Python, you will want to change the `rolling()` `mean()` to be other functions, such as `median()`, not the `groupby()` `mean()` that is part of the `agg()`.

Then plot the results to create Figure 7-3:

```
## Python
sns.scatterplot(draft_chart_py, x="Pick", y="roll_DrAV")
plt.show()
```

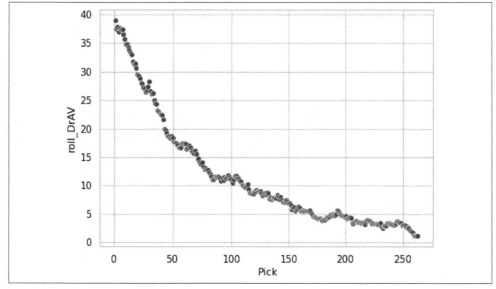

Figure 7-3. Scatterplot for draft pick number against draft approximate value (seaborn)

In R, `group_by()` pick and `summarize()` with the `mean()`. Replace NA values with 0. Then use the `rollapply()` function from the `zoo` package (make sure you have run `library(zoo)` and installed the package, since this is the first time you've used this package in the book). With `rollapply()`, use `width = 13`, and the `mean()` function (`FUN = mean`).

Tell the `mean()` function to ignore NA values with `na.rm = TRUE`. Fill in missing values with NA, `center` the mean (so that the rolling average is centered on the current DrAV), and calculate the mean when fewer than 13 observations are present (such as the start and end of the dataframe):

```
## R
draft_chart_r <-
    draft_r_use_pre2019 |>
    group_by(Pick) |>
    summarize(mean_DrAV = mean(DrAV, na.rm = TRUE)) |>
    mutate(mean_DrAV = ifelse(is.na(mean_DrAV),
                              0, mean_DrAV
    )) |>
    mutate(
        roll_DrAV =
            rollapply(mean_DrAV,
                width = 13,
                FUN = mean,
                na.rm = TRUE,
                fill = "extend",
                partial = TRUE
            )
    )
```

 For the exercise in R, you will want to change the `rollapply()` `mean()` to be other functions such as `median()`, not the `group_by()` `mean()` that is part of `summarize()`.

Then plot to create Figure 7-4:

```
## R
ggplot(draft_chart_r, aes(Pick, roll_DrAV)) +
    geom_point() +
    geom_smooth() +
    theme_bw() +
    ylab("Rolling average (\u00B1 6) DrAV") +
    xlab("Draft pick")
```

From here, you can simply fit a model to the data to help quantify this result. This model will allow you to use numbers rather than only examining a figure. You can use various models, and some of them (like LOESS curves or GAMs) are beyond the scope of this book. We have you fit a simple linear model to the logarithm of the data, while fixing the y-intercept—and transform back using an exponential function.

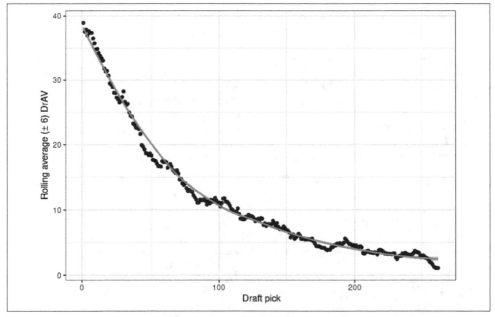

Figure 7-4. Scatterplot with smoothed trendline for draft pick number against draft approximate value (ggplot2)

 The log(0) function is mathematically undefined, so people will often add a value to allow this transformation to occur. Often a small number like 1 or 0.1 is used. Beware: this transformation can change your model results sometimes, so you may want to try different values for the number you add.

In Python, first drop the index (so you can access Pick with the model) and then plot:

```
## Python
draft_chart_py.reset_index(inplace=True)

draft_chart_py["roll_DrAV_log"] =\
    np.log(draft_chart_py["roll_DrAV"] + 1)

DrAV_pick_fit_py = \
    smf.ols(formula="roll_DrAV_log ~ Pick",
            data=draft_chart_py)\
    .fit()

print(DrAV_pick_fit_py.summary())
```

Resulting in:

```
                           OLS Regression Results
================================================================================
Dep. Variable:         roll_DrAV_log   R-squared:                      0.970
Model:                           OLS   Adj. R-squared:                 0.970
Method:                Least Squares   F-statistic:                    8497.
Date:               Sun, 04 Jun 2023   Prob (F-statistic):          1.38e-200
Time:                       09:42:13   Log-Likelihood:                177.05
No. Observations:                262   AIC:                           -350.1
Df Residuals:                    260   BIC:                           -343.0
Df Model:                          1
Covariance Type:           nonrobust
================================================================================
                 coef    std err          t      P>|t|      [0.025      0.975]
--------------------------------------------------------------------------------
Intercept      3.4871      0.015    227.712      0.000       3.457       3.517
Pick          -0.0093      0.000    -92.180      0.000      -0.010      -0.009
================================================================================
Omnibus:                       3.670   Durbin-Watson:                  0.101
Prob(Omnibus):                 0.160   Jarque-Bera (JB):               3.748
Skew:                          0.274   Prob(JB):                       0.154
Kurtosis:                      2.794   Cond. No.                       304.
================================================================================
```

Notes:
[1] Standard Errors assume that the covariance matrix of the errors is
    correctly specified.

And then merge back into draft_chart_py and look at the top of the data:

```python
## Python
draft_chart_py["fitted_DrAV"] = \
    np.exp(DrAV_pick_fit_py.predict()) - 1

draft_chart_py\
    .head()
```

Resulting in:

```
   Pick  DrAV_mean  roll_DrAV  roll_DrAV_log  fitted_DrAV
0     1      47.60  38.950000       3.687629    31.386918
1     2      39.85  37.575000       3.652604    31.086948
2     3      44.45  37.883333       3.660566    30.789757
3     4      31.15  36.990000       3.637323    30.495318
4     5      43.65  37.627273       3.653959    30.203606
```

In R, use the following:

```
## R
DrAV_pick_fit_r <-
    draft_chart_r |>
    lm(formula = log(roll_DrAV + 1) ~ Pick)

summary(DrAV_pick_fit_r)
```

Resulting in:

```
Call:
lm(formula = log(roll_DrAV + 1) ~ Pick, data = draft_chart_r)

Residuals:
     Min      1Q   Median      3Q     Max
-0.32443 -0.07818 -0.02338  0.08797  0.34123

Coefficients:
             Estimate Std. Error t value Pr(>|t|)
(Intercept)  3.4870598  0.0153134  227.71  <2e-16 ***
Pick        -0.0093052  0.0001009  -92.18  <2e-16 ***
---
Signif. codes:  0 '***' 0.001 '**' 0.01 '*' 0.05 '.' 0.1 ' ' 1

Residual standard error: 0.1236 on 260 degrees of freedom
Multiple R-squared:  0.9703,    Adjusted R-squared:  0.9702
F-statistic:  8497 on 1 and 260 DF,  p-value: < 2.2e-16
```

And then merge back into draft_chart_r and look at the top of the data:

```
## R
draft_chart_r <-
    draft_chart_r |>
    mutate(
        fitted_DrAV =
            pmax(
                0,
                exp(predict(DrAV_pick_fit_r)) - 1
            )
    )
draft_chart_r |>
    head()
```

Resulting in:

```
# A tibble: 6 × 4
   Pick mean_DrAV roll_DrAV fitted_DrAV
  <int>     <dbl>     <dbl>       <dbl>
1     1      47.6      39.0        31.4
2     2      39.8      37.6        31.1
3     3      44.4      37.9        30.8
4     4      31.2      37.0        30.5
5     5      43.6      37.6        30.2
6     6      34.7      37.4        29.9
```

So, to recap, in this section, you just calculated the estimated value for each pick. Notice that this fit likely underestimates the value of the pick at the very beginning of the draft because of the two-parameter nature of the exponential regression. This can be improved upon with a different model type, like the examples mentioned previously. The shortcomings notwithstanding, this estimate will allow you to explore draft situations, something you'll do in the very next section.

The Jets/Colts 2018 Trade Evaluated

Now that you have an estimate for the worth of each draft pick, let's look at what this model would have said about the trade between the Jets and the Colts in Table 7-1:

```
## R
library(kableExtra)
future_pick <-
    tibble(
        Pick = "Future 2nd round",
        Value = "14.8 (discounted at rate of 25%)"
    )

team <- tibble("Receiving team" = c("Jets", rep("Colts", 4)))

tbl_1 <-
    draft_chart_r |>
    filter(Pick %in% c(3, 6, 37, 49)) |>
    select(Pick, fitted_DrAV) |>
    rename(Value = fitted_DrAV) |>
    mutate(
        Pick = as.character(Pick),
        Value = as.character(round(Value, 1))
    ) |>
    bind_rows(future_pick)

team |>
    bind_cols(tbl_1) |>
    kbl(format = "pipe") |>
    kable_styling()
```

Table 7-1. Trade between the Jets and Colts

Receiving team	Pick	Value
Jets	3	30.8
Colts	6	29.9
Colts	37	22.2
Colts	49	19.7
Colts	Future 2nd round	14.8 (discounted at rate of 25%)

As you can see, the future pick is discounted at 25% because a rookie contract is four years and the waiting one year is a quarter, or 25%, of the contract for a current year.

Adding up the values from Table 7-1, it looks like the Jets got fleeced, losing an expected 55.6 DrAV in the trade. That's more than the value of the first-overall pick! These are just the statistically "expected" values from a generic draft pick at that position, using the model developed in this chapter based on previous drafts. Now, as of the writing of this book, what was predicted has come to fruition. You can use new data to see the actual DrAV for the players by creating Table 7-2:

```R
## R
library(kableExtra)

future_pick <-
    tibble(
        Pick = "Future 2nd round",
        Value = "14.8 (discounted at rate of 25)"
    )

results_trade <-
    tibble(
        Team = c("Jets", rep("Colts", 5)),
        Pick = c(
            3, 6, 37,
            "49-traded for 52",
            "49-traded for 169",
            "52 in 2019"
        ),
        Player = c(
            "Sam Darnold",
            "Quenton Nelson",
            "Braden Smith",
            "Kemoko Turay",
            "Jordan Wilkins",
            "Rock Ya-Sin"
        ),
        "DrAV" = c(25, 55, 32, 5, 8, 11)
    )
```

```
results_trade |>
    kbl(format = "pipe") |>
    kable_styling()
```

Table 7-2. Trade results between the Jets and Colts

Team	Pick	Player	DrAV
Jets	3	Sam Darnold	25
Colts	6	Quenton Nelson	55
Colts	37	Braden Smith	32
Colts	49—traded for 52	Kemoko Turay	5
Colts	49—traded for 169	Jordan Wilkins	8
Colts	52 in 2019	Rock Ya-Sin	11

So, the final tally was Jets 25 DrAV, Colts 111—a loss of 86 DrAV, which is almost three times the first overall pick!

This isn't always the case when a team trades up to draft a quarterback. For example, the Chiefs traded up for Patrick Mahomes, using two first-round picks and a third-round pick in 2017 to select the signal caller from Texas Tech, and that worked out to the tune of 85 DrAV, and (as of 2022) two Super Bowl championships.

More robust ways to price draft picks, can be found in the sources mentioned in "Analyzing the NFL Draft" on page 182. Generally speaking, using market-based data—such as the size of a player's first contract after their rookie deal—is the industry standard. Pro Football Reference's DrAV values are a decent proxy but have some issues; namely, they don't properly account for positional value—a quarterback is much, much more valuable if teams draft that position compared to any other position. For more on draft curves, *The Drafting Stage: Creating a Marketplace for NFL Draft Picks* (self-published, 2020) by Eric's former colleague at PFF, Brad Spielberger, and the founder of Over The Cap, Jason Fitzgerald, is a great place to start.

Are Some Teams Better at Drafting Players Than Others?

The question of whether some teams are better at drafting than others is a hard one because of the way in which draft picks are assigned to teams. The best teams choose at the end of each round, and as we've seen, the better players are picked before the weaker ones. So we could mistakenly assume that the worst teams are the best drafters, and vice versa. To account for this, we need to adjust expectations for each pick, using the model we created previously. Doing so, and taking the average and standard deviations of the difference between DrAV and fitted_DrAV, and aggregating over the 2000–2019 drafts, we arrive at the following ranking using Python:

```Python
## Python
draft_py_use_pre2019 = \
    draft_py_use_pre2019\
    .merge(draft_chart_py[["Pick", "fitted_DrAV"]],
        on="Pick")

draft_py_use_pre2019["OE"] = (
    draft_py_use_pre2019["DrAV"] -
    draft_py_use_pre2019["fitted_DrAV"]
)

draft_py_use_pre2019\
    .groupby("Tm")\
    .agg({"OE": ["count", "mean", "std"]})\
    .reset_index()\
    .sort_values([("OE", "mean")], ascending=False)
```

Resulting in:

	Tm	OE		
		count	mean	std
26	PIT	161	3.523873	18.878551
11	GNB	180	3.371433	20.063320
8	DAL	160	2.461129	16.620351
1	ATL	148	2.291654	16.124529
21	NOR	131	2.263655	18.036746
22	NWE	176	2.162438	20.822443
13	IND	162	1.852253	15.757658
4	CAR	148	1.842573	16.510813
2	BAL	170	1.721930	16.893993
27	SEA	181	1.480825	16.950089
16	LAC	144	1.393089	14.608528
5	CHI	149	0.672094	16.052031
20	MIN	167	0.544533	13.986365
15	KAN	154	0.501463	15.019527
25	PHI	162	0.472632	15.351785
6	CIN	176	0.466203	15.812953
14	JAX	158	0.182685	13.111672
30	TEN	172	0.128566	12.662670
12	HOU	145	-0.075827	12.978999
28	SFO	184	-0.092089	13.449491
31	WAS	150	-0.450485	9.951758
24	NYJ	137	-0.534640	13.317478
0	ARI	149	-0.601563	14.295335
23	NYG	145	-0.879900	12.471611
29	TAM	153	-0.922181	11.409698
3	BUF	161	-0.985761	12.458855
17	LAR	175	-1.439527	11.985219
19	MIA	151	-1.486282	10.470145
9	DEN	159	-1.491545	12.594449
10	DET	155	-1.765868	12.061696
18	LVR	162	-2.587423	10.217426
7	CLE	170	-3.557266	10.336729

The `.reset_index()` function in Python helps us because a dataframe in pandas has row names (an *index*) that can get confused when appending values.

Or in R:

```R
## R
draft_r_use_pre2019 <-
    draft_r_use_pre2019 |>
    left_join(draft_chart_r |> select(Pick, fitted_DrAV),
        by = "Pick"
    )

draft_r_use_pre2019 |>
    group_by(Tm) |>
    summarize(
        total_picks = n(),
        DrAV_OE = mean(DrAV - fitted_DrAV, na.rm = TRUE),
        DrAV_sigma = sd(DrAV - fitted_DrAV, na.rm = TRUE)
    ) |>
    arrange(-DrAV_OE) |>
    print(n = Inf)
```

Resulting in:

```
# A tibble: 32 × 4
   Tm    total_picks DrAV_OE DrAV_sigma
   <chr>       <int>   <dbl>      <dbl>
 1 PIT           161    3.52       18.9
 2 GNB           180    3.37       20.1
 3 DAL           160    2.46       16.6
 4 ATL           148    2.29       16.1
 5 NOR           131    2.26       18.0
 6 NWE           176    2.16       20.8
 7 IND           162    1.85       15.8
 8 CAR           148    1.84       16.5
 9 BAL           170    1.72       16.9
10 SEA           181    1.48       17.0
11 LAC           144    1.39       14.6
12 CHI           149   0.672       16.1
13 MIN           167   0.545       14.0
14 KAN           154   0.501       15.0
15 PHI           162   0.473       15.4
16 CIN           176   0.466       15.8
17 JAX           158   0.183       13.1
18 TEN           172   0.129       12.7
19 HOU           145 -0.0758       13.0
20 SFO           184 -0.0921       13.4
21 WAS           150  -0.450        9.95
22 NYJ           137  -0.535       13.3
23 ARI           149  -0.602       14.3
```

24	NYG	145	-0.880	12.5
25	TAM	153	-0.922	11.4
26	BUF	161	-0.986	12.5
27	LAR	175	-1.44	12.0
28	MIA	151	-1.49	10.5
29	DEN	159	-1.49	12.6
30	DET	155	-1.77	12.1
31	LVR	162	-2.59	10.2
32	CLE	170	-3.56	10.3

To no one's surprise, some of the storied franchises in the NFL have drafted the best above the draft curve since 2000: the Pittsburgh Steelers, the Green Bay Packers, and the Dallas Cowboys.

It also won't surprise anyone that last three teams on this list, the Cleveland Browns, the Oakland/Las Vegas Raiders, and the Detroit Lions, all have huge droughts in terms of team success as of the time of this writing. The Raiders haven't won their division since last playing in the Super Bowl in 2002, the Lions haven't won theirs since it was called the NFC Central in 1993, and the Cleveland Browns left the league, became the Baltimore Ravens, and came back since their last division title in 1989.

The question is, is this success and futility statistically significant? In Appendix B, we talk about standard errors and credible intervals. One reason we added the standard deviation to this table is so that we could easily compute the standard error for each team. This may be done in Python:

```Python
## Python
draft_py_use_pre2019 = \
    draft_py_use_pre2019\
    .merge(draft_chart_py[["Pick", "fitted_DrAV"]],
        on="Pick")

draft_py_use_pre2019_tm = (
    draft_py_use_pre2019.groupby("Tm")
    .agg({"OE": ["count", "mean", "std"]})
    .reset_index()
    .sort_values([("OE", "mean")], ascending=False)
)

draft_py_use_pre2019_tm.columns = \
    list(map("_".join, draft_py_use_pre2019_tm.columns))

draft_py_use_pre2019_tm.reset_index(inplace=True)

draft_py_use_pre2019_tm["se"] = (
    draft_py_use_pre2019_tm["OE_std"] /
    np.sqrt(draft_py_use_pre2019_tm["OE_count"])
)

draft_py_use_pre2019_tm["lower_bound"] = (
    draft_py_use_pre2019_tm["OE_mean"] - 1.96 * draft_py_use_pre2019_tm["se"]
```

```
)

draft_py_use_pre2019_tm["upper_bound"] = (
    draft_py_use_pre2019_tm["OE_mean"] + 1.96 * draft_py_use_pre2019_tm["se"]
)

print(draft_py_use_pre2019_tm)
```

Resulting in:

```
     index  Tm_  OE_count  ...         se  lower_bound  upper_bound
0       26  PIT       161  ...   1.487838     0.607710     6.440036
1       11  GNB       180  ...   1.495432     0.440387     6.302479
2        8  DAL       160  ...   1.313954    -0.114221     5.036479
3        1  ATL       148  ...   1.325428    -0.306186     4.889493
4       21  NOR       131  ...   1.575878    -0.825066     5.352375
5       22  NWE       176  ...   1.569551    -0.913882     5.238757
6       13  IND       162  ...   1.238039    -0.574302     4.278809
7        4  CAR       148  ...   1.357180    -0.817501     4.502647
8        2  BAL       170  ...   1.295710    -0.817661     4.261522
9       27  SEA       181  ...   1.259890    -0.988560     3.950210
10      16  LAC       144  ...   1.217377    -0.992970     3.779149
11       5  CHI       149  ...   1.315034    -1.905372     3.249560
12      20  MIN       167  ...   1.082297    -1.576770     2.665836
13      15  KAN       154  ...   1.210308    -1.870740     2.873667
14      25  PHI       162  ...   1.206150    -1.891423     2.836686
15       6  CIN       176  ...   1.191946    -1.870012     2.802417
16      14  JAX       158  ...   1.043109    -1.861808     2.227178
17      30  TEN       172  ...   0.965520    -1.763852     2.020984
18      12  HOU       145  ...   1.077847    -2.188407     2.036754
19      28  SFO       184  ...   0.991510    -2.035448     1.851270
20      31  WAS       150  ...   0.812558    -2.043098     1.142128
21      24  NYJ       137  ...   1.137789    -2.764706     1.695427
22       0  ARI       149  ...   1.171119    -2.896957     1.693831
23      23  NYG       145  ...   1.035711    -2.909893     1.150093
24      29  TAM       153  ...   0.922419    -2.730123     0.885761
25       3  BUF       161  ...   0.981895    -2.910275     0.938754
26      17  LAR       175  ...   0.905997    -3.215282     0.336228
27      19  MIA       151  ...   0.852048    -3.156297     0.183732
28       9  DEN       159  ...   0.998805    -3.449202     0.466113
29      10  DET       155  ...   0.968819    -3.664752     0.133017
30      18  LVR       162  ...   0.802757    -4.160827    -1.014020
31       7  CLE       170  ...   0.792791    -5.111136    -2.003396

[32 rows x 8 columns]
```

Or in R:

```
## R
draft_r_use_pre2019 |>
    group_by(Tm) |>
    summarize(
        total_picks = n(),
        DrAV_OE = mean(DrAV - fitted_DrAV,
```

```
          na.rm = TRUE
      ),
      DrAV_sigma = sd(DrAV - fitted_DrAV,
          na.rm = TRUE
      )
  ) |>
  mutate(
      se = DrAV_sigma / sqrt(total_picks),
      lower_bound = DrAV_OE - 1.96 * se,
      upper_bound = DrAV_OE + 1.96 * se
  ) |>
  arrange(-DrAV_OE) |>
  print(n = Inf)
```

Resulting in:

```
# A tibble: 32 × 7
   Tm    total_picks DrAV_OE DrAV_sigma    se lower_bound upper_bound
   <chr>       <int>   <dbl>      <dbl> <dbl>       <dbl>       <dbl>
 1 PIT           161    3.52       18.9  1.49       0.608        6.44
 2 GNB           180    3.37       20.1  1.50       0.440        6.30
 3 DAL           160    2.46       16.6  1.31      -0.114        5.04
 4 ATL           148    2.29       16.1  1.33      -0.306        4.89
 5 NOR           131    2.26       18.0  1.58      -0.825        5.35
 6 NWE           176    2.16       20.8  1.57      -0.914        5.24
 7 IND           162    1.85       15.8  1.24      -0.574        4.28
 8 CAR           148    1.84       16.5  1.36      -0.818        4.50
 9 BAL           170    1.72       16.9  1.30      -0.818        4.26
10 SEA           181    1.48       17.0  1.26      -0.989        3.95
11 LAC           144    1.39       14.6  1.22      -0.993        3.78
12 CHI           149   0.672       16.1  1.32      -1.91         3.25
13 MIN           167   0.545       14.0  1.08      -1.58         2.67
14 KAN           154   0.501       15.0  1.21      -1.87         2.87
15 PHI           162   0.473       15.4  1.21      -1.89         2.84
16 CIN           176   0.466       15.8  1.19      -1.87         2.80
17 JAX           158   0.183       13.1  1.04      -1.86         2.23
18 TEN           172   0.129       12.7 0.966      -1.76         2.02
19 HOU           145 -0.0758       13.0  1.08      -2.19         2.04
20 SFO           184 -0.0921       13.4 0.992      -2.04         1.85
21 WAS           150  -0.450        9.95 0.813     -2.04         1.14
22 NYJ           137  -0.535       13.3  1.14      -2.76         1.70
23 ARI           149  -0.602       14.3  1.17      -2.90         1.69
24 NYG           145  -0.880       12.5  1.04      -2.91         1.15
25 TAM           153  -0.922       11.4 0.922      -2.73        0.886
26 BUF           161  -0.986       12.5 0.982      -2.91        0.939
27 LAR           175   -1.44       12.0 0.906      -3.22        0.336
28 MIA           151   -1.49       10.5 0.852      -3.16        0.184
29 DEN           159   -1.49       12.6 0.999      -3.45        0.466
30 DET           155   -1.77       12.1 0.969      -3.66        0.133
31 LVR           162   -2.59       10.2 0.803      -4.16        -1.01
32 CLE           170   -3.56       10.3 0.793      -5.11        -2.00
```

Looking at this long code output, the 95% confidence intervals (CIs) can help you see which teams' DrAV_OE differed from 0. In both the Python and R outputs, 95% CIs are lower_bound and upper_bound. If this interval does not contain the value, you can consider it statistically different from 0. If the DrAV_OE is greater than the interval, the team did statistically better than average. If the DrAV_OE is less than the interval, the team did statistically worse than average.

So, using a 95% CI, it looks like two teams are statistically significantly better at drafting players than other teams, pick for pick (the Steelers and Packers), while two teams are statistically significantly worse at drafting players than the other teams (the Raiders and Browns). This is consistent with the research on the topic, which suggests that over a reasonable time interval (such as the average length of a general manager or coach's career), it's very hard to discern drafting talent.

The way to "win" the NFL Draft is to make draft pick trades like the Colts did against the Jets and give yourself more bites at the apple, as it were. Timo Riske discusses this more in the PFF article "A New Look at Historical Draft Success for all 32 NFL Teams" (*https://oreil.ly/-KdlX*).

One place where that famously failed, however, is with one of the two teams that has been historically bad at drafting since 2002, the Oakland/Las Vegas Raiders. In 2018, the Raiders traded their best player, edge player Khalil Mack, to the Chicago Bears for two first-round picks and an exchange of later-round picks. The Raiders were unable to ink a contract extension with Mack, whom the Bears later signed for the richest deal in the history of NFL defensive players. The Sloan Sports Analytics Conference (*https://oreil.ly/Ppf5a*)—the most high-profile gathering of sports analytics professionals in the world—lauded the trade for the Raiders, giving that team the award for the best transaction at the 2019 conference.

Generally speaking, trading one player for a bunch of players is going to go well for the team that acquires the bunch of players, even if they are draft picks. However, the Raiders, proven to be statistically notorious for bungling the picks, were unable to do much with the selections, with the best pick of the bunch being running back Josh Jacobs. Jacobs did lead the NFL in rushing yards in 2022, but prior to that point had failed to distinguish himself in the NFL, failing to earn a fifth year on his rookie contract. The other first-round pick in the trade, Damon Arnette, lasted less than two years with the team, while with Mack on the roster, the Bears made the playoffs twice and won a division title in his first year with the club in 2018.

Now, you've seen the basics of web scraping. What you do with this data is largely up to you! Like almost anything, the more you web scrape, the better you will become.

A suggestion for finding URLs is to use your web browser (such as Google Chrome, Microsoft Edge, or Firefox) inspection tool. This shows the HTML code for the web page you are visiting. You can use this code to help find the path for the table that you want based on the HTML and CSS selectors.

Data Science Tools Used in This Chapter

This chapter covered the following topics:

- Web scraping data in Python and R for the NFL Draft and NFL Scouting Combine data from Pro Football Reference (*https://www.pro-football-reference.com*)

- Using `for` loops in both Python and R

- Calculating rolling averages in Python with `rolling()` and in R with `rollapply()`

- Reapplying data-wrangling tools you learned in previous chapters

Exercises

1. Change the web-scraping examples to different ranges of years for the NFL Draft. Does an error come up? Why?

2. Use the process laid out in this chapter to scrape NFL Scouting Combine data by using the general URL *https://www.pro-football-reference.com/draft/YEAR-combine.htm* (you will need to change *YEAR*). This is a preview for Chapter 8, where you will dive into this data further.

3. With NFL Scouting Combine data, plot the 40-yard dash times for each player, with point color determined by the position the player plays. Which position is the fastest? The slowest? Do the same for the other events. What are the patterns you see?

4. Use the NFL Scouting Combine data in conjunction with the NFL Draft data scraped in this chapter. What is the relationship between where a player is selected in the NFL Draft and their 40-yard dash time? Is this relationship more pronounced for some positions? How does your answer to question 3 influence your approach to this question?

5. For the draft curve exercise, change the six-pick moving average to a six-pick moving median. What happens? Is this preferable?

Suggested Readings

Many books and other resources exist on web scraping. Besides the package documentation for `rvest` in R and `read_html()` in `pandas`, here are two good ones to start with:

- *R Web Scraping Quick Start Guide* by Olgun Aydin (Packt Publishing, 2018)
- *Web Scraping with Python,* 2nd edition, by Ryan Mitchell (O'Reilly, 2018); 3rd edition forthcoming in 2024

The Drafting Stage: Creating a Marketplace for NFL Draft Picks, referenced previously in this chapter, provides an overview of the NFL Draft along with many great details.

## CHAPTER 8

Principal Component Analysis and Clustering: Player Attributes

In this era of big data, some people have a strong urge to simply "throw the kitchen sink" at data in an attempt to find patterns and draw value from them. In football analytics, this urge is strong as well. This approach should generally be taken with caution, because of the dynamic and small-sampled nature of the game. But if handled with care, the process of *unsupervised learning* (in contrast to *supervised learning*, both of which are defined in a couple of paragraphs) can yield insights that are useful to us as football analysts.

In this chapter, you will use NFL Scouting Combine data from 2000 to 2023 that you will obtain through Pro Football Reference. As mentioned in Chapter 7, the NFL Scouting Combine is a yearly event, usually held in Indianapolis, Indiana, where NFL players go through a battery of physical (and other) tests in preparation for the NFL Draft. The entire football industry gets together for what is essentially its yearly conference, and with many of the top players no longer testing while they are there (opting to test at friendlier *Pro Days* held at their college campuses), the importance of the on-field events in the eyes of many have waned. Furthermore, the addition of tracking data into our lives as analysts has given rise to more accurate (and timely) estimates of player athleticism, which is a fast-rising set of problems in and of its own.

Several recent articles provide discussion on the NFL Scouting Combine. "Using Tracking and Charting Data to Better Evaluate NFL Players: A Review" (*https://oreil.ly/cJd5R*) by Eric and others from the Sloan Sports Analytics Conference discusses other ways to measure player data. "Beyond the 40-Yard Dash: How Player Tracking Could Modernize the NFL Combine" (*https://oreil.ly/M67dm*) by Sam Fortier of the *Washington Post* describes how player tracking data might be used. "Inside the Rams' Major Changes to their Draft Process, and Why They Won't Go Back to 'Normal'" (*https://oreil.ly/Us5oh*) by Jourdan Rodrigue of the *Athletic* provides coverage of the Rams' drafting process.

The eventual obsolete nature of (at least the on-field testing portion of) the NFL Scouting Combine notwithstanding, the data provides a good vehicle to study both principal component analysis and clustering. *Principal component analysis* (PCA) is the process of taking a set of features that possess collinearity of some form (*collinearity* in this context means the predictors conceptually contain duplicative information and are numerically correlated) and "mushing" them into smaller subsets of features that are each (linearly) independent of one another.

Athleticism data is chock-full of these sorts of things. For example, how fast a player runs the 40-yard dash depends very much on how much they weigh (although not perfectly), whereas how high they jump is correlated with how far they can jump. Being able to take a set of features that are all-important in their own right, but not completely independent of one another, and creating a smaller set of data column-wise, is a process that is often referred to as *dimensionality reduction* in data science and related fields.

Clustering is the process of dividing data points into similar groups (*clusters*) based on a set of features. We can cluster data in many ways, but if the groups are not known a priori, this process falls into the category of unsupervised learning. *Supervised learning* algorithms require a predefined response variable to be trained on. In contrast, *unsupervised learning* essentially allows the data to create the response variable—which in the case of clustering is the cluster—out of thin air.

Clustering is a really effective approach in team and individual sports because players are often grouped, either formally or informally, into position groups in team sports. Sometimes the nature of these position groups changes over time, and data can help us detect that change in an effort to help teams adjust their process of building rosters with players that fit into their ideas of positions.

Additionally, in both team and individual sports, players have *styles* that can be uncovered in the data. While a set of features depicting a player's style—often displayed through some sort of plot—can be helpful to the mathematically inclined, traditionalists often want to have player types described by groups; hence the value of clustering here as well.

You can imagine that the process of running a PCA in the data before clustering is essential. If how fast a player runs the 40-yard dash carries much of the same signal as how high they jump when tested for the vertical jump, treating them both as one variable (while not "double counting" in the traditional sense) will be overcounting some traits in favor of others. Thus, you'll start with the process of running a PCA on data, which is adapted from Eric's DataCamp course (*https://oreil.ly/RRNpW*).

We present basic, introductory methods for multivariate statistics in this chapter. Advanced methods emerge on a regular basis. For example, uniform manifold approximation and projection is emerging as one popular new distance-based tool at the time of this writing, and topological data analysis (which uses geometric properties rather than distance) is another. If you understand the basic methods that we present, learning these new tools will be easier, and you will have a benchmark by which to compare these new methods.

Web Scraping and Visualizing NFL Scouting Combine Data

To obtain the data, you'll use similar web-scraping tools as in Chapter 7. Note the change from draft to combine in the code for the URL. Also, the data requires some cleaning. Sometimes the data contains extra headings. To remove these, remove rows whose value equals the heading (such as Ht != "Ht"). In both languages, height (Ht) needs to be converted from foot-inch to inches.

With Python, use this code to download and save the data:

```
## Python
import pandas as pd
import seaborn as sns
import matplotlib.pyplot as plt

combine_py = pd.DataFrame()
for i in range(2000, 2023 + 1):
    url = (
            "https://www.pro-football-reference.com/draft/" +
            str(i) +
            "-combine.htm"
    )
    web_data = pd.read_html(url)[0]
    web_data["Season"] = i
    web_data = web_data.query('Ht != "Ht"')
    combine_py = pd.concat([combine_py, web_data])

combine_py.reset_index(drop=True, inplace=True)
combine_py.to_csv("combine_data_py.csv", index=False)

combine_py[["Ht-ft", "Ht-in"]] = \
    combine_py["Ht"].str.split("-", expand=True)

combine_py = \
    combine_py\
    .astype({
        "Wt": float,
        "40yd": float,
        "Vertical": float,
        "Bench": float,
```

```
        "Broad Jump": float,
        "3Cone": float,
        "Shuttle": float,
        "Ht-ft": float,
        "Ht-in": float
        })

combine_py["Ht"] = (
    combine_py["Ht-ft"] * 12.0 +
    combine_py["Ht-in"]
)

combine_py\
    .drop(["Ht-ft", "Ht-in"], axis=1, inplace=True)

combine_py.describe()

              Ht            Wt  ...     Shuttle       Season
count  7970.000000  7975.000000  ...  4993.000000  7999.000000
mean     73.801255   242.550094  ...     4.400925  2011.698087
std       2.646040    45.296794  ...     0.266781     6.950760
min      64.000000   144.000000  ...     3.730000  2000.000000
25%      72.000000   205.000000  ...     4.200000  2006.000000
50%      74.000000   232.000000  ...     4.360000  2012.000000
75%      76.000000   279.500000  ...     4.560000  2018.000000
max      82.000000   384.000000  ...     5.560000  2023.000000

[8 rows x 9 columns]
```

With R, use this code (in R, you will save the data and read the data back into to R because this is a quick way to help R recognize the column types):

```
## R
library(tidyverse)
library(rvest)
library(htmlTable)
library(multiUS)
library(ggthemes)

combine_r <- tibble()
for (i in seq(from = 2000, to = 2023)) {
    url <- paste0("https://www.pro-football-reference.com/draft/",
                  i,
                  "-combine.htm")
    web_data <-
        read_html(url) |>
        html_table()
    web_data_clean <-
        web_data[[1]] |>
        mutate(Season = i) |>
        filter(Ht != "Ht")
    combine_r <-
        bind_rows(combine_r,
```

```
                web_data_clean)
}
write_csv(combine_r, "combine_data_r.csv")
combine_r <- read_csv("combine_data_r.csv")

combine_r <-
    combine_r |>
    mutate(ht_ft = as.numeric(str_sub(Ht, 1, 1)),
           ht_in = str_sub(Ht, 2, 4),
           ht_in = as.numeric(str_remove(ht_in, "-")),
           Ht = ht_ft * 12 + ht_in) |>
    select(-Ht_ft, -Ht_in)
summary(combine_r)
```

Player	Pos	School	College
Length:7999	Length:7999	Length:7999	Length:7999
Class :character	Class :character	Class :character	Class :character
Mode :character	Mode :character	Mode :character	Mode :character

Ht	Wt	40yd	Vertical	Bench
Min. :64.0	Min. :144.0	Min. :4.220	Min. :17.50	Min. : 2.00
1st Qu.:72.0	1st Qu.:205.0	1st Qu.:4.530	1st Qu.:30.00	1st Qu.:16.00
Median :74.0	Median :232.0	Median :4.690	Median :33.00	Median :21.00
Mean :73.8	Mean :242.6	Mean :4.774	Mean :32.93	Mean :20.74
3rd Qu.:76.0	3rd Qu.:279.5	3rd Qu.:4.970	3rd Qu.:36.00	3rd Qu.:25.00
Max. :82.0	Max. :384.0	Max. :6.050	Max. :46.50	Max. :49.00
NA's :29	NA's :24	NA's :583	NA's :1837	NA's :2802

Broad Jump	3Cone	Shuttle	Drafted (tm/rnd/yr)
Min. : 74.0	Min. :6.280	Min. :3.730	Length:7999
1st Qu.:109.0	1st Qu.:6.980	1st Qu.:4.200	Class :character
Median :116.0	Median :7.190	Median :4.360	Mode :character
Mean :114.8	Mean :7.285	Mean :4.401	
3rd Qu.:121.0	3rd Qu.:7.530	3rd Qu.:4.560	
Max. :147.0	Max. :9.120	Max. :5.560	
NA's :1913	NA's :3126	NA's :3006	

Season
Min.   :2000
1st Qu.:2006
Median :2012
Mean   :2012
3rd Qu.:2018
Max.   :2023

Notice here that many, many players don't have a full set of data, as evidenced by the NA's in the R table. A lot of players just give their heights and weights at the combine, which is why the number of NAs there are few. You will resolve this issue in a bit, but let's look at the reason you need PCA in the first place: the pair-wise correlations between events.

First, look at height versus weight. In Python, create Figure 8-1:

```python
# Python
sns.set_theme(style="whitegrid", palette="colorblind")

sns.regplot(data=combine_py, x="Ht", y="Wt");
plt.show();
```

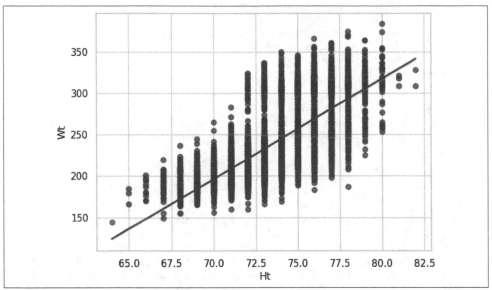

Figure 8-1. Scatterplot with trendline for player height plotted against player weight, plotted with seaborn

Or in R, create Figure 8-2:

```r
# R
ggplot(combine_r, aes(x = Ht, y = Wt)) +
    geom_point() +
    theme_bw() +
    xlab("Player Height (inches)") +
    ylab("Player Weight (pounds)") +
    geom_smooth(method = "lm", formula = y ~ x)
```

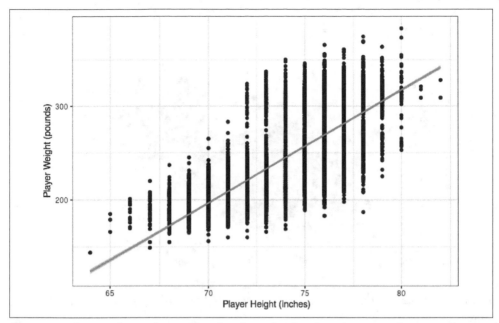

Figure 8-2. Scatterplot with trendline for player height plotted against player weight, plotted with ggplot2

This makes sense: the taller you are, the more you weigh. Hence, there really aren't two independent pieces of information here. Let's look at weight versus 40-yard dash. In Python, create Figure 8-3:

```
# Python
sns.regplot(data=combine_py,
            x="Wt",
            y="40yd",
            line_kws={"color": "red"});
plt.show();
```

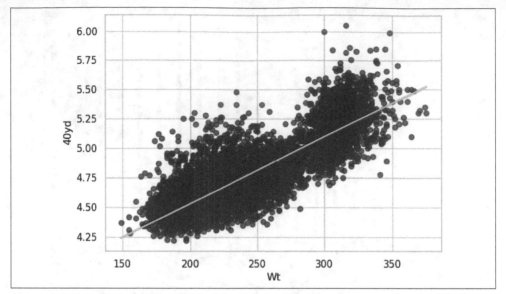

Figure 8-3. Scatterplot with trendline for player weight plotted against 40-yard dash time (seaborn)

Or in R, create a similar figure (Figure 8-4):

```
# R
ggplot(combine_r, aes(x = Wt, y = `40yd`)) +
    geom_point() +
    theme_bw() +
    xlab("Player Weight (pounds)") +
    ylab("Player 40-yard dash (seconds)") +
    geom_smooth(method = "lm", formula = y ~ x)
```

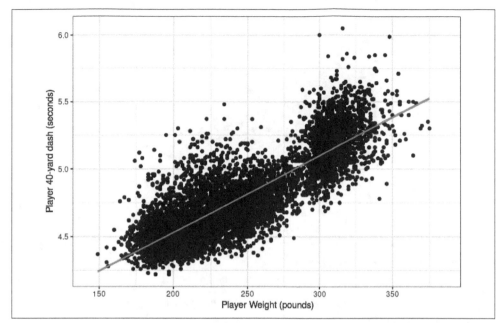

Figure 8-4. Scatterplot with trendline for player weight plotted against 40-yard dash time (ggplot2)

 In most computer languages, starting object names with numbers, like 40yd, is bad. The computer does not know what to do because the computer thinks something like arithmetic is about to happen but then the computer gets some letters instead. R lets you use improper names by placing backticks (`) around them, as you did to create Figure 8-3.

Here, again, you have a positive correlation. Notice here that two clusters are emerging already in the data, though: a lot of really heavy players (above 300 lbs) and a lot of really light players (near 225 lbs). These two groups serve as an example of a bimodal distribution: rather than having a single center like a normal distribution, two groups exist. Now, let's look at 40-yard dash and vertical jump. In Python, create Figure 8-5:

```
# Python
sns.regplot(data=combine_py,
            x="40yd",
            y="Vertical",
            line_kws={"color": "red"});
plt.show();
```

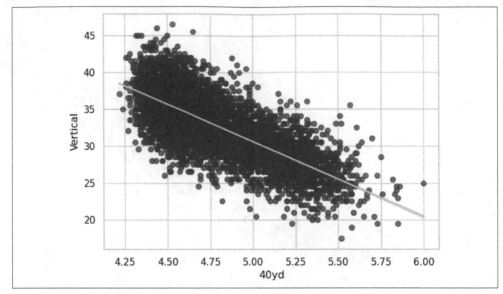

Figure 8-5. Scatterplot with trendline for player 40-yard dash time plotted against vertical jump (seaborn)

Or in R, create Figure 8-6:

```
# R
ggplot(combine_r, aes(x = `40yd`, y = Vertical)) +
    geom_point() +
    theme_bw() +
    xlab("Player 40-yard dash (seconds)") +
    ylab("Player vertical jump (inches)") +
    geom_smooth(method = "lm", formula = y ~ x)
```

Resulting in:

```
geom_smooth: na.rm = FALSE, orientation = NA, se = TRUE
stat_smooth: na.rm = FALSE, orientation = NA, se = TRUE, method = lm,
formula = y ~ x position_identity
```

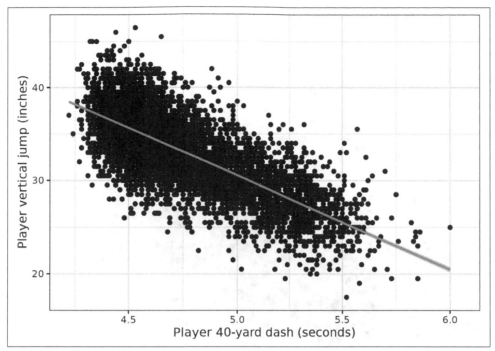

Figure 8-6. Scatterplot with trendline for player 40-yard dash time plotted against vertical jump (ggplot2)

Here, you have a negative relationship; the faster the player (lower the 40-yard dash, in seconds), the higher the vertical jump (in inches). Does agility (as measured by the three-cone drill, see Figure 8-7) also track with the 40-yard dash?

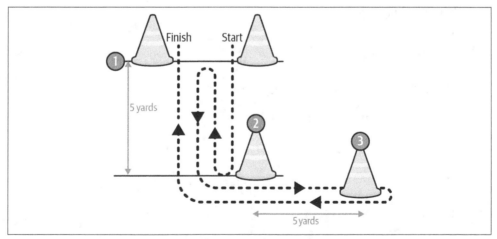

Figure 8-7. In the three-cone drill, players run around the cones, following the path, and their time is recorded

In Python, create Figure 8-8:

```python
# Python
sns.regplot(data=combine_py,
            x="40yd",
            y="3Cone",
            line_kws={"color": "red"});
plt.show();
```

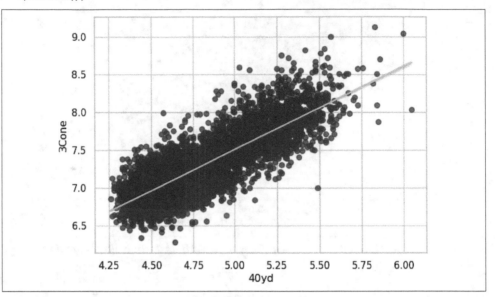

Figure 8-8. Scatterplot with trendline for player 40-yard dash time plotted against their three-cone drill (seaborn)

Or in R, create Figure 8-9:

```r
# R
ggplot(combine_r, aes(x = `40yd`, y = `3Cone`)) +
    geom_point() +
    theme_bw() +
    xlab("Player 40-yard dash (seconds)") +
    ylab("Player 3 cone drill (inches)") +
    geom_smooth(method = "lm", formula = y ~ x)
```

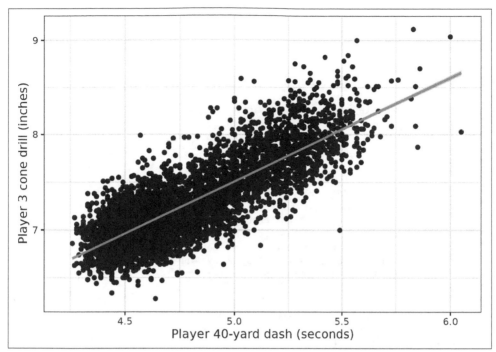

Figure 8-9. Scatterplot with trend line for player 40-yard dash time plotted against three-cone drill time (*ggplot2*)

We have another positive relationship. Hence, it's clear that athleticism is measured in many ways, and it's reasonable to assume that none of them are independent of the others.

To commence the process of PCA, we first have to "fill in" the missing data. We use k-nearest neighbors, which is beyond the scope of this book. As you learn more about your data structure, you may want to do your own research to find other methods for replacing missing values. For now, just run the code.

 We included the web-scraping step (again) and imputation step so that the book would be self-contained and you can recreate all our data yourself.

With this method, only players with a recorded height and weight at the combine had their data *imputed* (that is, missing values were estimated using a statistical method). We have included an `if-else` statement so this code will run only if the file has not been downloaded and saved to the current directory.

In Python, run the following:

```python
## Python
import numpy as np
import os
from sklearn.impute import KNNImputer

combine_knn_py_file = "combine_knn_py.csv"
col_impute = ["Ht", "Wt", "40yd", "Vertical",
              "Bench", "Broad Jump", "3Cone",
              "Shuttle"]

if not os.path.isfile(combine_knn_py_file):
    combine_knn_py = combine_py.drop(col_impute, axis=1)
    imputer = KNNImputer(n_neighbors=10)
    knn_out_py = imputer.fit_transform(combine_py[col_impute])
    knn_out_py = pd.DataFrame(knn_out_py)
    knn_out_py.columns = col_impute
    combine_knn_py = pd.concat([combine_knn_py, knn_out_py], axis=1)
    combine_knn_py.to_csv(combine_knn_py_file)

else:
    combine_knn_py = pd.read_csv(combine_knn_py_file)

combine_knn_py.describe()
```

Resulting in:

```
         Unnamed: 0        Season  ...         3Cone       Shuttle
count   7999.000000   7999.000000  ...   7999.000000   7999.000000
mean    3999.000000   2011.698087  ...      7.239512      4.373727
std     2309.256735      6.950760  ...      0.374693      0.240294
min        0.000000   2000.000000  ...      6.280000      3.730000
25%     1999.500000   2006.000000  ...      6.978000      4.210000
50%     3999.000000   2012.000000  ...      7.122000      4.310000
75%     5998.500000   2018.000000  ...      7.450000      4.510000
max     7998.000000   2023.000000  ...      9.120000      5.560000

[8 rows x 10 columns]
```

In R, run this:

```r
## R
combine_knn_r_file <- "combine_knn_r.csv"

if (!file.exists(combine_knn_r_file)) {
    imput_input <-
        combine_r |>
        select(Ht:Shuttle) |>
        as.data.frame()

    knn_out_r <-
        KNNimp(imput_input, k = 10,
               scale = TRUE,
```

```
            meth = "median") |>
        as_tibble()

    combine_knn_r <-
        combine_r |>
        select(Player:College, Season) |>
        bind_cols(knn_out_r)
    write_csv(x = combine_knn_r,
              file = combine_knn_r_file)
} else {
    combine_knn_r <- read_csv(combine_knn_r_file)
}

combine_knn_r |>
    summary()
```

Resulting in:

```
    Player                Pos                School              College
 Length:7999         Length:7999         Length:7999         Length:7999
 Class :character    Class :character    Class :character    Class :character
 Mode  :character    Mode  :character    Mode  :character    Mode  :character



     Season              Ht               Wt                40yd            Vertical
 Min.   :2000      Min.   :64.0      Min.   :144.0     Min.   :4.22     Min.   :17.50
 1st Qu.:2006      1st Qu.:72.0      1st Qu.:205.0     1st Qu.:4.53     1st Qu.:30.50
 Median :2012      Median :74.0      Median :232.0     Median :4.68     Median :33.50
 Mean   :2012      Mean   :73.8      Mean   :242.5     Mean   :4.77     Mean   :32.93
 3rd Qu.:2018      3rd Qu.:76.0      3rd Qu.:279.0     3rd Qu.:4.97     3rd Qu.:35.50
 Max.   :2023      Max.   :82.0      Max.   :384.0     Max.   :6.05     Max.   :46.50
     Bench           Broad Jump          3Cone            Shuttle
 Min.   : 2.00     Min.   : 74.0     Min.   :6.280     Min.   :3.730
 1st Qu.:16.00     1st Qu.:109.0     1st Qu.:6.975     1st Qu.:4.210
 Median :19.50     Median :116.0     Median :7.140     Median :4.330
 Mean   :20.04     Mean   :114.7     Mean   :7.252     Mean   :4.383
 3rd Qu.:24.00     3rd Qu.:121.0     3rd Qu.:7.470     3rd Qu.:4.525
 Max.   :49.00     Max.   :147.0     Max.   :9.120     Max.   :5.560
```

Notice, no more missing data. But, for reasons that are beyond this book and similar to reasons explained in this Stack Overflow post (*https://oreil.ly/-SoJi*), the two methods produce similar, but slightly different, results because they use slightly different assumptions and methods.

Introduction to PCA

Before you fit the PCA used for later analysis, let's take a short break to look at how a PCA works. Conceptually, a PCA reduces the number of dimensions of data to use the fewest possible. Graphically, *dimensions* refers to the number of axes needed to describe the data. Tabularly, *dimensions* refers to the number of columns needed to describe the data. Algebraically, *dimensions* refers to the number of independent variables needed to describe the data.

Look back at the length-weight relation shown in Figures 8-1 and 8-2. Do we need an x-axis and a y-axis to describe this data? Probably not. Let's fit a PCA, and then you can look at the outputs. You will use the raw data after removing missing values. In Python, use the PCA() function from the scikit-learn package:

```Python
## Python
from sklearn.decomposition import PCA

pca_wt_ht = PCA(svd_solver="full")
wt_ht_py = \
    combine_py[["Wt", "Ht"]]\
    .query("Wt.notnull() & Ht.notnull()")\
    .copy()

pca_fit_wt_ht_py = \
    pca_wt_ht.fit_transform(wt_ht_py)
```

Or with R, use the prcomp() function from the stats package that is part of the core set of R packages:

```r
## r
wt_ht_r <-
    combine_r |>
    select(Wt, Ht) |>
    filter(!is.na(Wt) & !is.na(Ht))

pca_fit_wt_ht_r <-
    prcomp(wt_ht_r)
```

Now, let's look at the model details. In Python, use this code to look at the variance explained by each of the new *principal components* (*PCs*), or new axes of data:

```Python
## Python
print(pca_wt_ht.explained_variance_ratio_)
```

Resulting in:

```
[0.99829949 0.00170051]
```

In R, look at the summary() of the fit model:

```
## R
summary(pca_fit_wt_ht_r)
```

Resulting in:

```
Importance of components:
                          PC1     PC2
Standard deviation     45.3195 1.8704
Proportion of Variance  0.9983 0.0017
Cumulative Proportion   0.9983 1.0000
```

Both of the outputs show that PC1 (or the axis of data) contains 99.8% of the variability within the data. This tells you that only the first PC is important for the data.

To see the new representation of the data, plot the new data. In Python, use matplot lib's plot() to create a simple scatterplot of the outputs to create Figure 8-10:

```Python
## Python
plt.plot(pca_fit_wt_ht_py[:, 0], pca_fit_wt_ht_py[:, 1], "o");
plt.show();
```

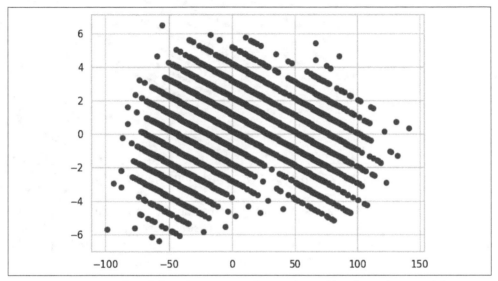

Figure 8-10. Scatterplot of the PCA rotation for weight and height with plot() from matplotlot

In R, use ggplot to create Figure 8-11:

```Python
## Python
pca_fit_wt_ht_r$x |>
    as_tibble() |>
    ggplot(aes(x = PC1, y = PC2)) +
    geom_point() +
    theme_bw()
```

A lot is going on in these figures. First, compare Figure 8-1 to Figure 8-10 or compare Figure 8-2 to Figure 8-11. If you are good at spatial pattern recognition, you might see that the figures have the same data, just rotated (try looking at the outliers that fall away from the main crowd, and you might be able to see how these data points moved). That's because PCA rotates data to create fewer dimensions of it. Because this example has only two dimensions, the pattern is visible in the data.

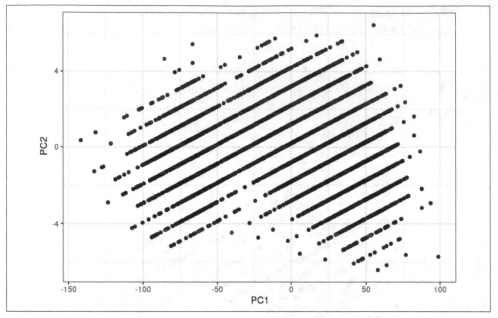

Figure 8-11. Scatterplot of the PCA rotation for weight and height with ggplot2

You can get these rotation values by looking at components_ in Python or by printing the fitted PCA in R. Although you will not use these values more in this book, they have uses in machine learning, when people need to convert data in and out of the PCA space. In Python, use this code to extract the rotation values:

```
## Python
pca_wt_ht.components_
```

Resulting in:

```
array([[ 0.9991454 ,  0.04133366],
       [-0.04133366,  0.9991454 ]])
```

In R, use this code to extract the rotation values:

```
## R
print(pca_fit_wt_ht_r)
```

Resulting in:

```
Standard deviations (1, .., p=2):
[1] 45.319497  1.870442

Rotation (n x k) = (2 x 2):
          PC1          PC2
Wt -0.99914540 -0.04133366
Ht -0.04133366  0.99914540
```

Your values may be different signs than our values (for example, we may have –0.999, and you may have 0.999), but that's OK and varies randomly. For example, Python's PCA has three positive numbers with the book's numbers, whereas R's PCA has the same numbers, but negative. These numbers say that you take a player's weight and multiply it by 0.999 and add 0.041 times the player's height to create the first PC.

Why are these values so different? Look back again at Figures 8-10 and 8-11. Notice that the x-axis and y-axis have vastly different scales. This can also cause different input features to have different levels of influence. Also, unequal numbers sometimes cause computational problems.

In the next section, you will put all features on the same unit level by scaling the inputs. *Scaling* refers to transforming a feature, usually to have a mean of 0 and standard deviation of 1. Thus, the different units and magnitudes of the features no longer are important after scaling.

PCA on All Data

Now, apply R and Python's built-in algorithms to perform the PCA analysis on all data. This will help you to create new, and fewer, predictor variables that are independent of one another. First, scale that data and then run the PCA. In Python:

```
## Python
from sklearn.decomposition import PCA

scaled_combine_knn_py = (
    combine_knn_py[col_impute] -
    combine_knn_py[col_impute].mean()) / \
    combine_knn_py[col_impute].std()

pca = PCA(svd_solver="full")
pca_fit_py = \
    pca.fit_transform(scaled_combine_knn_py)
```

Or in R:

```
## R
scaled_combine_knn_r <-
    scale(combine_knn_r |> select(Ht:Shuttle))

pca_fit_r <-
    prcomp(scaled_combine_knn_r)
```

The object `pca_fit` is more of a model object than it is a data object. It has some interesting nuggets. For one, you can look at the weights for each of the PCs. In Python:

```
## Python
rotation = pd.DataFrame(pca.components_, index=col_impute)
print(rotation)
```

Resulting in:

```
                 0         1          2  ...         5          6          7
Ht         0.280591  0.393341   0.390341  ...  -0.367237   0.381342   0.377843
Wt         0.506953  0.273279  -0.063500  ...   0.359464  -0.110215  -0.130109
40yd      -0.709435 -0.001356  -0.082813  ...  -0.096306   0.068683  -0.019891
Vertical  -0.203781  0.033044   0.012393  ...   0.296674   0.523851   0.509379
Bench     -0.142324  0.161150   0.593645  ...  -0.369323  -0.035026  -0.428910
Broad Jump 0.206559 -0.080594  -0.613440  ...  -0.641948   0.277464  -0.094715
3Cone     -0.005106 -0.044482   0.027751  ...  -0.298070  -0.677284   0.620678
Shuttle   -0.237684  0.857359  -0.327257  ...   0.035926  -0.162377  -0.047622

[8 rows x 8 columns]
```

Or in R:

```
## R
print(pca_fit_r$rotation)
```

Resulting in:

```
                 PC1         PC2          PC3          PC4          PC5
Ht        -0.2797884   0.4656585   0.747620897   0.21562254  -0.06128240
Wt        -0.3906321   0.2803488   0.002635803  -0.04180851   0.14466721
40yd      -0.3937993  -0.0994878   0.045495814  -0.01403113   0.48636319
Vertical   0.3456004   0.4186011   0.002756780  -0.53609337   0.54959455
Bench     -0.2668254   0.6109690  -0.642865102   0.18678232  -0.16950424
Broad Jump  0.3674448  0.3388903   0.139855673  -0.31774247  -0.43822503
3Cone     -0.3823998  -0.1115644  -0.060381468  -0.51566448   0.04602633
Shuttle   -0.3770497  -0.1373520   0.049975069  -0.51225797  -0.46240812
                 PC6         PC7          PC8
Ht         0.1394355  -0.164300591  -0.22194154
Wt        -0.1553924   0.089682873   0.84494838
40yd      -0.4083595   0.543622221  -0.36595878
Vertical   0.3348374   0.043881290  -0.04282478
Bench      0.0604662  -0.009981142  -0.27362455
Broad Jump -0.6246519  0.216613922  -0.02156424
3Cone     -0.2978993  -0.675688165  -0.15604839
Shuttle    0.4415283   0.404889985  -0.03684955
```

 PCA components are based on a mathematical property of matrices called *eigenvalues* and *eigenvectors*. These are scalar, and your PCs might have opposite signs compared to our examples (for example, if our Ht for PC1 is negative, yours might be positive). Not to worry. Just note that is why the signs are different. Also, this appears to have occurred with R and Python when we were writing this book.

Notice that the first PC weighs 40-yard dash and weight about the same (with a factor weight of –0.39). Notice that most of the nonsize metrics that are better when smaller have a negative weight (40-yard dash, the agility drills), while those that are better when bigger (vertical and broad jump) are positive.

 Our Python and R examples start to diverge slightly because of differences in the imputation methods and PCA across the two languages. However, the qualitative results are the same.

This is a good sign that you're onto something. You can look at how much of the proportion of variance is explained by each PC. In Python:

```python
## Python
print(pca.explained_variance_)
```

Resulting in:

```
[5.60561713 0.83096684 0.62448842 0.37527929 0.21709371 0.13913206
 0.12108346 0.08633909]
```

Or in R, look at the standard deviation squared:

```r
## R
print(pca_fit_r$sdev^2)
```

Which results in:

```
[1] 5.67454385 0.84556662 0.61894619 0.35175168 0.19651463 0.11815058 0.11166060
[8] 0.08286586
```

Notice here that, as expected, the first PC handles a significant amount of the variability in the data, but subsequent PCs have some influence over it as well. If you take these standard deviations in R, convert them to variances by squaring them (such as $PCA1^2$) and then dividing by the sum of all variances, you can see the percent variance explained by each axis. Python's PCA includes this without extra math.

In Python:

```python
## Python
pca_percent_py = \
    pca.explained_variance_ratio_.round(4) * 100
print(pca_percent_py)
```

Resulting in:

```
[70.07 10.39  7.81  4.69  2.71  1.74  1.51  1.08]
```

Or in R:

```r
## R
pca_var_r <- pca_fit_r$sdev^2
pca_percent_r <-
    round(pca_var_r / sum(pca_var_r) * 100, 2)
print(pca_percent_r)
```

Resulting in:

```
[1] 70.93 10.57  7.74  4.40  2.46  1.48  1.40  1.04
```

Now, to access the actual PCs, deploy the following code to get something you can more readily use. In Python:

```
## Python
pca_fit_py = pd.DataFrame(pca_fit_py)
pca_fit_py.columns = \
    ["PC" + str(x + 1) for x in range(len(pca_fit_py.columns))]

combine_knn_py = \
    pd.concat([combine_knn_py, pca_fit_py], axis=1)
```

Or in R:

```
## R
combine_knn_r <-
    combine_knn_r |>
    bind_cols(pca_fit_r$x)
```

The first thing to do from here is graph the first few PCs to see if you have any natural clusters emerging. In Python, create Figure 8-12:

```
## Python
sns.scatterplot(data=combine_knn_py,
                x="PC1",
                y="PC2");
plt.show();
```

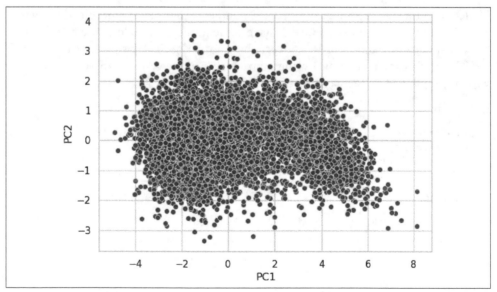

Figure 8-12. Plot of first two PCA components (seaborn)

In R, create Figure 8-13:

```R
## R
ggplot(combine_knn_r, aes(x = PC1, y = PC2)) +
    geom_point() +
    theme_bw() +
    xlab(paste0("PC1 = ", pca_percent_r[1], "%")) +
    ylab(paste0("PC2 = ", pca_percent_r[2], "%"))
```

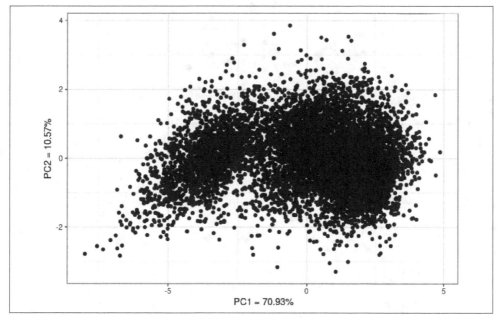

Figure 8-13. Plot of first two PCA components (`ggplot2`)

 Because PCs are based on eigenvalues, your figures might be flipped from each of our examples. For example, our R and Python figures are mirror images of each other.

You can already see two clusters! What if you use more of the data to reveal other possibilities? Shade each of the points by the value of the third PC. In Python, create Figure 8-14:

```Python
## Python
sns.scatterplot(data=combine_knn_py,
                x="PC1",
                y="PC2",
                hue="PC3");
plt.show();
```

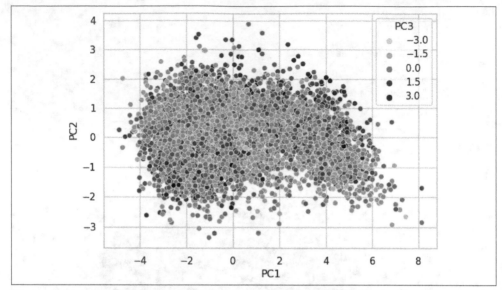

Figure 8-14. Plot of the first two PCA components, with the third PCA component as the point color (seaborn)

In R, create Figure 8-15:

```
## R
ggplot(combine_knn_r,
       aes(x = PC1, y = PC2, color = PC3)) +
   geom_point() +
   theme_bw() +
   xlab(paste0("PC1 = ", pca_percent_r[1], "%")) +
   ylab(paste0("PC2 = ", pca_percent_r[2], "%")) +
   scale_color_continuous(
       paste0("PC3 = ", pca_percent_r[3], "%"),
       low="skyblue", high="navyblue")
```

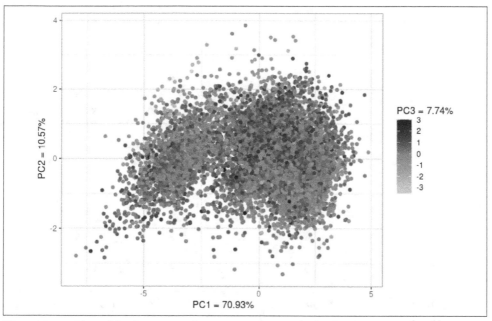

Figure 8-15. Plot of the first two PCA components, with the third PCA component as the point color (ggplot2)

Interesting. Looks like players on the edges of the plot have lower values of PC3, corresponding to darker shades. What if you shade by position?

In Python, create Figure 8-16:

```
## Python
sns.scatterplot(data=combine_knn_py,
                x="PC1",
                y="PC2",
                hue="Pos");
plt.show();
```

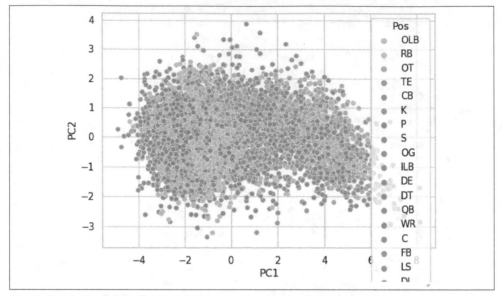

Figure 8-16. Plot of the first two PCA components, with the point player position as the color (seaborn)

In R, use a colorblind-friendly palette to create Figure 8-17:

```R
## R
library(RColorBrewer)
color_count <- length(unique(combine_knn_r$Pos))
get_palette <- colorRampPalette(brewer.pal(9, "Set1"))

ggplot(combine_knn_r,
       aes(x = PC1, y = PC2, color = Pos)) +
    geom_point(alpha = 0.75) +
    theme_bw() +
    xlab(paste0("PC1 = ", pca_percent_r[1], "%")) +
    ylab(paste0("PC2 = ", pca_percent_r[2], "%")) +
    scale_color_manual("Player position",
                       values = get_palette(color_count))
```

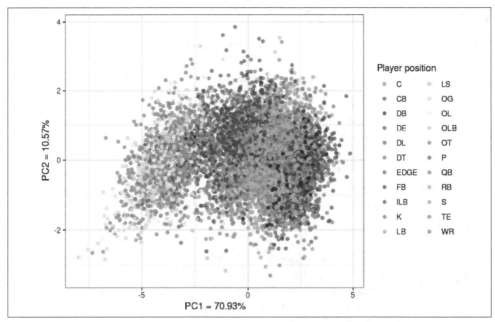

Figure 8-17. Plot of the first two PCA components, with the point player position as the color (ggplot2)

OK, this is fun. It does look like the positions create clear groupings in the data. Upon first blush, it looks like you can split this data into about five to seven clusters.

 About 8% of men and 0.5% of women in the US have *color vision deficiency* (more commonly known as *colorblindness*). This ranges from seeing only black-and-white to, more commonly, not being able to tell all colors apart. For example, Richard has trouble with reds and greens, which also makes purple hard for him to see. Hence, try to pick colors more people can use. Tools such as Color Oracle (*https://www.colororacle.org*) and Sim Daltonism (*https://michelf.ca/projects/sim-daltonism*) let you test your figures and see them like a person with colorblindness.

Clustering Combine Data

The clustering algorithm you're going to use here is k-means clustering, which aims to partition a dataset into *k* clusters in which each observation belongs to the cluster with the nearest mean (*cluster centers*, or *cluster centroid*), serving as a prototype of the cluster.

We break this section into two units because the numerical methods diverge. This does not mean either method is "wrong" or "better" than the other. Instead, this helps you see that methods can simply be different. Also, understanding and interpreting multivariate statistical methods (especially unsupervised methods) can be subjective. One of Richard's professors at Texas Tech, Stephen Cox, would describe this process as similar to reading tea leaves because you can often find whatever pattern you are looking for if you are not careful.

 Despite our comparison of multivariate statistics being similar to reading tea leaves, the approaches are commonly used (and rightly so) because the methods are powerful and useful. You, as a user and modeler, however, need to understand this limitation if you are going to use the methods.

Clustering Combine Data in Python

 If you are coding along (and hopefully you are!), your results will likely be different from ours for the clustering. You will need to look at the results to see which cluster number corresponds to the groups on your computer.

To start with Python, use `kmeans` from the `scipy` package and fit for six centers (we set `seed` to be 1234 so that we would get the same results each time we ran the code for the book):

```
## Python
from scipy.cluster.vq import vq, kmeans

k_means_fit_py = \
    kmeans(combine_knn_py[["PC1", "PC2"]], 6, seed = 1234)
```

Next, attach these clusters to the dataset:

```
## Python
combine_knn_py["cluster"] = \
    vq(combine_knn_py[["PC1", "PC2"]], k_means_fit_py[0])[0]

combine_knn_py.head()
```

Resulting in:

```
    Unnamed: 0            Player  Pos  ...       PC7       PC8 cluster
0            0      John Abraham  OLB  ... -0.146522  0.292073       3
1            1    Shaun Alexander   RB  ... -0.073008  0.060237       1
2            2     Darnell Alford   OT  ... -0.491523 -0.068370       0
3            3      Kyle Allamon   TE  ...  0.328718 -0.059768       2
4            4  Rashard Anderson   CB  ... -0.674786 -0.276374       1

[5 rows x 24 columns]
```

Not much can be gleaned from the head of the data here. However, one thing you can do is see if the clusters bring like positions and player types together. Look at cluster 1:

```
print(
    combine_knn_py.query("cluster == 1")
    .groupby("Pos")
    .agg({"Ht": ["count", "mean"], "Wt": ["count", "mean"]})
)
```

Resulting in:

	Ht		Wt	
	count	mean	count	mean
Pos				
CB	219	72.442922	219	197.506849
DB	27	72.074074	27	201.074074
DE	13	75.384615	13	250.307692
EDGE	5	74.800000	5	247.400000
FB	8	72.500000	8	235.000000
ILB	20	73.700000	20	237.700000
K	9	73.777778	9	207.666667
LB	45	73.673333	45	234.486667
OLB	78	73.987179	78	236.397436
P	24	74.416667	24	206.041667
QB	40	74.450000	40	217.000000
RB	163	71.225767	163	216.932515
S	221	72.800905	221	209.330317
TE	20	75.450000	20	244.100000
WR	409	73.795844	409	207.871638

You have little bit of everything here, but it's mostly players away from the ball: cornerbacks, safeties, and wide receivers, with some running backs mixed in. Among those position groups, these players are heavier players.

Depending on the random-number generator in your computer, different clusters will have different numbers, so be wary of comparing across clustering regimes. Now, let's look at the summary for all clusters by using a plot. In Python, create Figure 8-18:

```
## Python
combine_knn_py_cluster = \
    combine_knn_py\
```

```
        .groupby(["cluster", "Pos"])\
        .agg({"Ht": ["count", "mean"],
              "Wt": ["mean"]}
)

combine_knn_py_cluster.columns = \
    list(map("_".join, combine_knn_py_cluster.columns))

combine_knn_py_cluster.reset_index(inplace=True)

combine_knn_py_cluster\
    .rename(columns={"Ht_count": "n",
                     "Ht_mean": "Ht",
                     "Wt_mean": "Wt"},
            inplace=True)

combine_knn_py_cluster.cluster = \
    combine_knn_py_cluster.cluster.astype(str)

sns.catplot(combine_knn_py_cluster, x="n", y="Pos",
            col="cluster", col_wrap=3, kind="bar");
plt.show();
```

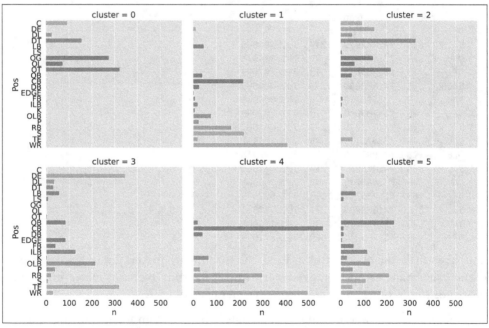

Figure 8-18. Plot of positions by cluster (seaborn)

Here, cluster 0 is largely bigger players, like offensive linemen and interior defensive linemen. Cluster 2 includes similar positions as cluster 0, while adding defensive ends and tight ends, which are also represented in great measure in cluster 3, along with outside linebackers. We talked about cluster 1 previously. Cluster 4 has a lot of the same positions as cluster 1, but more defensive backs and wide receivers (we'll take a look at size next). Cluster 5 includes a lot of quarterbacks, as well as a decent number of other skill positions (*skill positions* in football are those that typically hold the ball and are responsible for scoring).

Let's look at the summary by cluster to compare weight and height:

```Python
## Python
combine_knn_py_cluster\
    .groupby("cluster")\
    .agg({"Ht": ["mean"], "Wt": ["mean"]})
```

Resulting in:

```
             Ht          Wt
           mean        mean
cluster
0      75.866972  293.708339
1      73.631939  223.254368
2      74.966517  272.490225
3      75.230958  250.847219
4      71.099940  205.290840
5      73.098379  229.605847
```

As we hypothesized, cluster 1 and cluster 4, while having largely the same positions represented, are such that cluster 1 includes much bigger players in terms of height and weight. Clusters 0 and 2 show similar results.

Clustering Combine Data in R

To start with R, use the `kmeans()` function that comes in the `stats` package with R's core packages. Here, `iter.max` is the maximum number of iterations allowed to find the clusters and centers in the number of clusters. This is needed because the algorithm requires multiple attempts, or *iterations*, to fit the model. This is accomplished using the following script (you'll set R's random seed with `set.seed(123)` so that you get consistent results):

```R
## R
set.seed(123)
k_means_fit_r <-
    kmeans(combine_knn_r |> select(PC1, PC2),
           centers = 6, iter.max = 10)
```

Next, attach these clusters to the dataset:

```R
## R
combine_knn_r <-
    combine_knn_r |>
    mutate(cluster = k_means_fit_r$cluster)
```

```
combine_knn_r |>
    select(Pos, Ht:Shuttle, cluster) |>
    head()
```

Resulting in:

```
# A tibble: 6 × 10
   Pos      Ht     Wt  `40yd` Vertical Bench `Broad Jump`  `3Cone` Shuttle cluster
   <chr> <dbl>  <dbl>   <dbl>    <dbl> <dbl>        <dbl>    <dbl>   <dbl>   <int>
1  OLB      76    252    4.55     38.5  23.5          124     6.94    4.22       6
2  RB       72    218    4.58     35.5  19            120     7.07    4.24       4
3  OT       76    334    5.56     25    23             94     8.48    4.98       3
4  TE       74    253    4.97     29    21            104     7.29    4.49       1
5  CB       74    206    4.55     34    15            123     7.18    4.15       4
6  K        70    202    4.55     36    16            120.    6.94    4.17       4
```

As with the Python example, not much can be gleaned from the head of the data here, other than the fact that in R the index begins at 1, rather than 0. Look at the first cluster:

```
## R
combine_knn_r |>
    filter(cluster == 1) |>
    group_by(Pos) |>
    summarize(n = n(), Ht = mean(Ht), Wt = mean(Wt)) |>
    arrange(-n) |>
    print(n = Inf)
```

Resulting in:

```
# A tibble: 21 × 4
   Pos       n    Ht    Wt
   <chr> <int> <dbl> <dbl>
1  QB      236  74.8  223.
2  TE      200  76.2  255.
3  DE      193  75.3  266.
4  ILB     127  73.0  242.
5  OLB     116  73.5  242.
6  FB       65  72.3  247.
7  P        60  74.8  221.
8  RB       49  71.1  226.
9  LB       43  72.9  235.
10 DT       38  73.9  288.
11 WR       29  74.9  219.
12 LS       28  73.9  241.
13 EDGE     27  75.3  255.
14 K        23  73.2  213.
15 DL       22  75.5  267.
16 S        11  72.6  220.
17 C         3  74.7  283
18 OG        2  75.5  300
19 CB        1  73    214
20 DB        1  72    197
21 OL        1  72    238
```

This cluster has its biggest representation among quarterbacks, tight ends, and defensive ends. This makes some sense football-wise, as many tight ends are converted quarterbacks, like former Vikings tight end (and head coach) Mike Tice, who was a quarterback at the University of Maryland before becoming a professional tight end.

Depending on the random-number generator in your computer, different clusters will have different numbers: In R, create Figure 8-19:

```R
## R
combine_knn_r_cluster <-
    combine_knn_r |>
    group_by(cluster, Pos) |>
    summarize(n = n(), Ht = mean(Ht), Wt = mean(Wt),
            .groups="drop")

combine_knn_r_cluster |>
    ggplot(aes(x = n, y = Pos)) +
    geom_col(position='dodge') +
    theme_bw() +
    facet_wrap(vars(cluster)) +
    theme(strip.background = element_blank()) +
    ylab("Position") +
    xlab("Count")
```

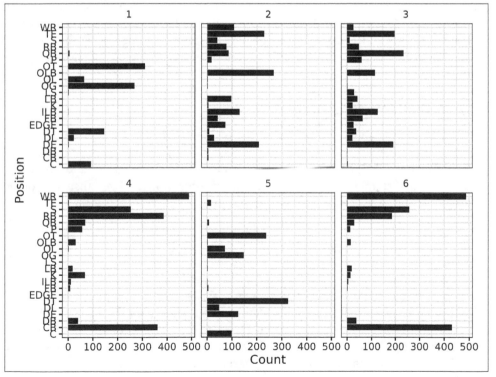

Figure 8-19. Plot of positions by cluster (ggplot2)

Here you get a similar result as previously with Python, with offensive and defensive linemen grouped together in some clusters, while skill-position players find other clusters more frequently. Let's look at the summary by cluster to compare weight and height. In R, use the following:

```
## R
combine_knn_r_cluster |>
    group_by(cluster) |>
    summarize(ave_ht = mean(Ht),
              ave_wt = mean(Wt))
```

Resulting in:

```
# A tibble: 6 × 3
  cluster ave_ht ave_wt
    <int>  <dbl>  <dbl>
1       1   73.8   242.
2       2   72.4   214.
3       3   75.6   291.
4       4   71.7   211.
5       5   75.7   281.
6       6   75.0   246.
```

Clusters 2 and 4 include players farther away from the ball and hence smaller, while clusters 3 and 5 are bigger players playing along the line of scrimmage. Clusters 1 (described previously) and 6 have more "tweener" athletes who play positions like quarterback, tight end, outside linebacker, and in some cases, defensive end (*tweener* players are those who can play multiple positions well but may or may not excel and be the best at any position).

Closing Thoughts on Clustering

Even this initial analysis shows the power of this approach, as after some minor adjustments to the data, you're able to produce reasonable groupings of the players without having to define the clusters beforehand. For players without a ton of initial data, this can get the conversation started vis-à-vis comparables, fits, and other things. It can also weed out players who do not fit a particular coach's scheme.

You can drill down even further once the initial groupings are made. Among just wide receivers, is this wide receiver a taller/slower type who wins on contested catches? Is he a shorter/shiftier player who wins with separation? Is he a unicorn of the mold of Randy Moss or Calvin Johnson? Does he make players who are already on the roster redundant—older players whose salaries the team might want to get rid of—or does he supplement the other players as a replacement for a player who has left through free agency or a trade? Arif Hasan of *The Athletic* discusses these traits for a specific coach as an example in "Vikings Combine Trends: What Might They Look For in Their Offensive Draftees?" (*https://oreil.ly/7zaxH*).

Clustering has been used on more advanced problems to group things like a receiver's pass routes. In this problem, you're using model-based curve clustering on the actual (*x,y*) trajectories of the players to do with math what companies like PFF have been doing with their eyes for some time: chart each play for analysis. As we've mentioned, most old-school coaches and front-office members are in favor of groupings, so methods like these will always have appeal in American football for that reason. Dani Chu and collaborators describe approaches such as route identification in "Route Identification in the National Football League" (*https://oreil.ly/BPi2e*), which also exists as an open-access preprint (*https://oreil.ly/OLwup*).

Some of the drawbacks of k-means clustering specifically are that it's very sensitive to initial conditions and to the number of clusters chosen. You can take steps—including random-number seeding (as done in the previous example)—to help reduce these issues, but a wise analyst or statistician understands the key assumptions of their methods. As with everything, vigilance is required as new data comes in, so that outputs are not too drastically altered with each passing year. Some years, because of evolution in the game, you might have to add or delete a cluster, but this decision should be made after thorough and careful analysis of the downstream effects of doing so.

Data Science Tools Used in This Chapter

This chapter covered the following topics:

- Adapting web-scraping tools from Chapter 7 for a slightly different web page
- Using PCA to reduce the number of dimensions
- Using cluster analysis to examine data for groups

Exercises

1. What happens when you do PCA on the original data, with all of the NAs included? How does this affect your future football analytics workflow?

2. Perform the k-means clustering in this chapter with the first three PCs. Do you see any differences? The first four?

3. Perform the k-means clustering in this chapter with five and seven clusters. How does this change the results?

4. What other problem in this book would be enhanced with a clustering approach?

Suggested Readings

Many basic statistics books cover the methods presented in this chapter. Here are two:

- *Essential Math for Data Science* by Thomas Nield (O'Reilly, 2022) provides a gentle introduction to statistics as well as mathematics for applied data scientists.
- *R for Data Science*, 2nd edition, by Hadley Wickham et al. (O'Reilly Media, 2023) provides an introduction to many tools and methods for applied data scientists.

## CHAPTER 9

Advanced Tools and Next Steps

This book has focused on the basics of football analytics using Python and R. We personally use both on a regular basis. However, we also use tools beyond these two programming languages. For people who want to keep growing, you will need to leave your comfort zone. This chapter provides an overview of other tools we use. We start with modeling tools that we use but have not mentioned yet, either because the topics are too advanced or because we could not find public data that would easily allow you to code along.

We then move on to computer tools. The topics are both disjointed and interwoven at the same time: you can learn one skill independently, but often, using one skill works best with other skills. As a football comparison, a linebacker needs to be able to defend the run, rush the passer, and cover players who are running pass routes, often in the same series of a game. Some skills (such as the ability to read a play) and player traits (such as speed) will help with all three linebacker situations, but they are often drilled separately. The most valuable players are great at all three.

This chapter is based on our experiences working as data scientists as well as an article Richard wrote for natural resource managers ("Paths to Computational Fluency for Natural Resource Educators, Researchers, and Managers" (*https://oreil.ly/oNokn*)). We suggest you learn the topics in the order we present them, and we list reasons in Table 9-1. Once you gain some comfort with a skill, move on to another area. Eventually, you'll make it back to a skill area and see where to grow in that area. As you learn more technologies, you can become better at learning new technologies!

 In *Build a Career in Data Science* (Manning, 2020), Emily Robinson and Jacqueline Nolis provide broader coverage of skills for a career in data science.

Table 9-1. Advanced tools, our reasons for using them, and example products

Tool	Reason	Examples
Command line	Efficiently and automatically work with your operating system, use other tools that are command-line-only.	Microsoft PowerShell, bash, Zsh
Version control	Keep track of changes to code, collaborate on code, share and publish code.	Git, Apache Subversion (SVN), Mercurial
Linting	Clean code, provide internal consistency for style, reduce errors, and improve quality.	Pylint, lintr, Black
Package creation and hosting	Reuse your own code, share your code internally or externally, and more easily maintain your code.	Conda, pip, CRAN
Environment	Provide reproducible results, take models to production or the cloud, and ensure the same tools across collaborations.	Conda, Docker, Poetry
Interactives and reports	Allow other people to explore data without knowing how to code, prototype tools before handing off to DevOps.	Jupyter Notebook, Shiny, Quarto
Cloud	Deploy tools and advanced computing resources, share data.	Amazon Web Services (AWS), Microsoft Azure, Google Cloud Platform (GCP)

All of the advanced tools we mention have free documentation either included with them or online. However, finding and using this documentation can be hard and often requires locating the proverbial diamonds in the rough. Hence, paid tutorials and resources such as books often, but not always, offer you a quality product. If you are a broke grad student, you might want to spend your time diving through the free resources to find the gems. If you are a working professional with kids and not much time, you probably want to pay for learning resources. Basically, finding quality learning materials comes down to time versus money.

Advanced Modeling Tools

Within this book, we have covered a wide range of models. For many people, these tools will be enough to advance your football analytics game. However, other people will want to go further and push the envelope. In this section, we describe some methods we use on a regular basis. Notice that many of these topics are interwoven. Hence, learning one topic might lead you to learn about another topic as well.

Time Series Analysis

Football data, especially feature-rich, high-resolution data within games, lends itself to look at trends through time. *Time series analysis* estimates trends through time. The methods are commonly used in finance, with applications in other fields such as ecology, physics, and social sciences. Basically, these models can provide better estimations when past observations are important for future predictions (also known as *auto-correlations*). Here are some resources we've found helpful:

- *Time Series Analysis and Its Application*, 4th edition, by Robert H. Shumway and David S. Stoffer (Springer, 2017). Provides a detailed introduction to time series analysis using R.

- *Practical Time Series Analysis* by Aileen Nielsen (O'Reilly, 2019) provides a gentler introduction to time series analysis with a focus on application, especially to machine learning.

- Prophet (*https://oreil.ly/nPbvF*) by Facebook's Core Data Science team is a time-series modeling tool that can be powerful when used correctly.

Multivariate Statistics Beyond PCA

Chapter 8 briefly introduced multivariate methods such as PCA and clustering. These two methods are the tip of the iceberg. Other methods exist, such as redundancy analysis (RDA), that allow both multivariate predictor and response variables. These methods form the basis of many entry-level unsupervised learning methods because the methods find their own predictive groups. Additionally, PCA assumes *Euclidean distance* (the same distance you may or may not remember from the Pythagorean theorem; for example, in two dimensions $c = \sqrt{\left(a^2 + b^2\right)}$). Other types of distances exist, and multivariate methods cover these. Lastly, many classification methods exist. For example, some multivariate methods extend to time series analysis, such as dynamic factor analysis (DFA).

Beyond direct application of these tools, understanding these methods will give you a firm foundation if you want to learn machine learning tools. Some books we learned from or think would be helpful include the following:

- *Numerical Ecology*, 3rd edition, by Pierre Legendre and Louis Legendre (Elsevier, 2012) provides a generally accessible overview of many multivariate methods.

- *Analysis of Multivariate Time Series Using the MARSS Package* vignette by E. Holmes, M. Scheuerell, and E. Ward comes with the MARSS package and may be found on the MARSS CRAN page (*https://oreil.ly/SQ2qZ*). This detailed introduction describes how to do time series analysis with R on multivariate data.

Quantile Regression

Usually, regression models the average (or mean) expected value. *Quantile regression* models other parts of a distribution—specifically, a user-specified quantile. Whereas a boxplot (covered in "Boxplots" on page 35) has predefined quantiles, the user specifies which quantile they want with quantile regression. For example, when looking at NFL Scouting Combine data, you might wonder how player speeds change through time. A traditional multiple regression would look at the average player through time. A quantile regression would help you see if the faster players get faster through time. Quantile regression would also be helpful when looking at the NFL Draft data in Chapter 7. Resources for learning about quantile regression include the package documentation:

- The `quantreg` package (*https://oreil.ly/diL03*) in R has a well-written 21-page vignette (*https://oreil.ly/P4ZVs*) that is great regardless of language.

- The `statsmodels` package in Python also has quantile regression documentation (*https://oreil.ly/0ha7Z*).

Bayesian Statistics and Hierarchical Models

Up to this point, our entire use of probability has been built on the assumption that events occur based on long-term occurrence, or frequency of events occurring. This type of probability is known as *frequentist* statistics. However, other views of probabilities exist.

Notably, a Bayesian perspective views the world in terms of degrees of belief or certainty. For example, a frequentist 95% confidence interval (CI) around a mean contains the mean 95% of the time, if you repeat your observations many, many, many times. Conversely, a Bayesian 95% credible interval (CrI) indicates a range for which you are 95% certain to contain the mean. It's a subtle, but important, difference.

A Bayesian perspective begins with a prior understanding of the system, updates that understanding by using observed data, and then generates a posterior distribution. In practice, Bayesian methods offer three major advantages:

- They can fit more complicated models when other methods might not have enough data.

- They can include multiple sources of information more readily.

- A Bayesian view of statistics is what many people have, even if they do not know the name for it.

For example, consider picking which team will win. The prior information can either come from other data, your best guess, or any other source. If you are humble, you might think you will be right 50% of the time and would guess you would get two games right and two games wrong. If you are overconfident, you might think you will be right 80% of the time and get eight games right and two games wrong. If you are underconfident, you might think you will be right 20% of the time and get eight games wrong and two games right. This is your prior distribution.

For this example, a beta distribution gives you the probability distribution given the number of "successes" and "failures." Graphically, this gives you Figure 9-1.

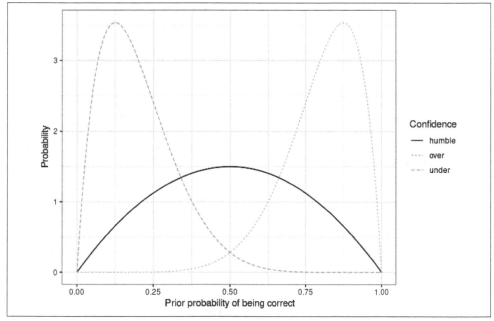

Figure 9-1. Prior distribution for predicting results of games

After observing 50 games, perhaps you were correct for 30 games and wrong for 20 games. A frequentist would say you are correct 60% of the time (30/50). To a Bayesian, this is the observed likelihood. With a Beta distribution, this would be 60 successes and 40 failures. Figure 9-2 shows the likelihood probability.

Next, a Bayesian multiplies Figure 9-1 by Figure 9-2 to create the posterior Figure 9-3. All three guesses are close, but the prior distribution informs the posterior.

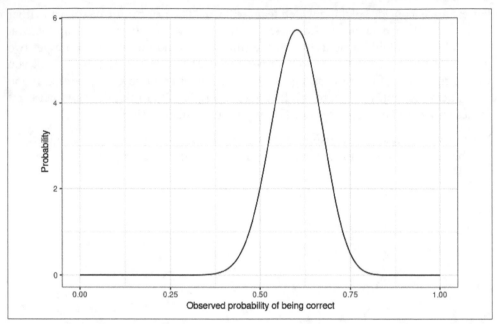

Figure 9-2. Likelihood distribution for predicting results of games

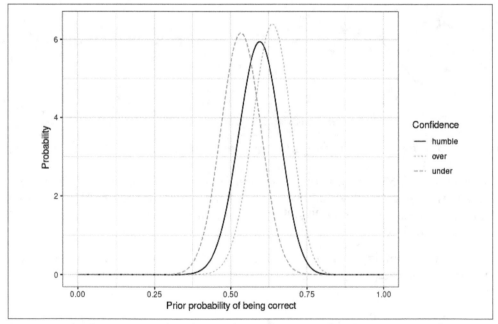

Figure 9-3. In this posterior distribution for predicting results of games, notice the influence of the prior distribution on the posterior

This simple example illustrates how Bayesian methods work for an easy problem. However, Bayesian models also allow much more complicated models, such as multi-level models, to be fit (for example, to examine a regression with both team-level and player-level features). Additionally, Bayesian models' posterior distribution captures uncertainty not also present with other estimation methods. For those of you wanting to know more about thinking like a Bayesian or doing Bayesian statistics, here are some books we have found to be helpful:

- *The Theory That Would Not Die: How Bayes' Rule Cracked the Enigma Code, Hunted Down Russian Submarines, and Emerged Triumphant from Two Centuries of Controversy* by Sharon Bertsch McGrayne (Yale University Press, 2011) describes how people have used Bayesian statistics to make decisions through high-profile examples such as the US Navy searching for missing nuclear weapons and submarines.

- *The Foundations of Statistics* by Leonard J. Savage (Dover Press, 1972) provides an overview of how to think like a Bayesian, especially in the context of decision making such as betting or management.

- *Doing Bayesian Data Analysis*, 2nd edition, by John Kruschke (Elsevier, 2014) is also known as the *puppy book* because of its cover. This book provides a gentle introduction to Bayesian statistics.

- *Bayesian Data Analysis*, 3rd edition, by Andrew Gelman et al. (CRC Press, 2013). This book, often called *BDA3* by Stan users, provides a rigorous and detailed coverage of Bayesian methods. Richard could not read this book until he had taken two years of advanced undergraduate and intro graduate-level math courses.

- *Statistical Rethinking: A Bayesian Course with Examples in R and Stan*, 2nd edition, by Richard McElreath (CRC Press, 2020). This book is between the puppy book and BDA3 in rigor and is an intermediate-level text for people wanting to learn Bayesian statistics.

Survival Analysis/Time-to-Event

How long does a quarterback last in the pocket until he either throws the ball or is sacked? *Time-to-event* or *survival* analysis would help you answer that question. We did not cover this technique in the book because we could not find public data for this analysis. However, for people with more detailed time data, this analysis would help you understand how long until events occur. Some books we found useful on this topic include the following:

- *Regression Modeling Strategies: With Applications to Linear Models, Logistic and Ordinal Regression, and Survival Analysis*, 2nd edition, by Frank E. Harrell Jr. (Springer, 2015). Besides being useful for regression, this book also includes survival analysis.

- *Think Stats*, 2nd edition, by Allen B. Downey (O'Reilly, 2014) includes an accessible chapter on survival analysis using Python.

Bayesian Networks/Structural Equation Modeling

Chapter 8 alluded to the interconnectedness of data. Taking this a step further, sometimes data has no clear-cut cause or effect, or cause and effect variables are linked. For example, consider combine draft attributes. A player's weight might be linked to a player's running speed (for example, lighter players run quicker). Running speed and weight might both be associated with a running back's rushing yards.

How to tease apart these confounding variables? Tools such as structural equation modeling and Bayesian networks allow these relations to be estimated. Here are some books we found to be helpful:

- *The Book of Why* by Judea Pearl and Dana Mackenzie (Basic Books, 2018) walks through how to think about the world in terms of networks. The book also provides a great conceptual introduction to network models.
- *Bayesian Networks With Examples in R*, 2nd edition, by Marco Scutari and Jean-Baptiste Denis (CRC Press, 2021) provides a nice introduction to Bayesian networks.
- *Structural Equation Modeling and Natural Systems* by James B. Grace (Cambridge University Press, 2006) provides a gentle introduction to these models using ecological data.

Machine Learning

Machine learning is not any single tool but rather a collection of tools and a method of thinking about data. Most of our book has focused on statistical understanding of data. In contrast, machine learning thinks about how to use data to make predictions in an automated fashion.

Many great books exist on this topic, but we do not have any strong recommendations. Instead, build a solid foundation in math, stats, and programming, and then you should be well equipped to understand.

Command Line Tools

Command lines allow you to use code to interact with computers. Command lines have several related names, that, although having specific technical definitions, are often used interchangeably. One is *shell*, because this is the outside, or "shell," of the operating system that humans, such as yourself, touch. Another is *terminal* because this is the software that uses the input and output text (historically, *terminal* referred to the hardware you used). As a more modern definition, this can refer to the software as well; for example, Richard's Linux computer calls his command-line application the *terminal*. Lastly, *console* refers to the physical terminal. The Ask Ubuntu site (*https://oreil.ly/ask*) provides detailed discussion on this topic, as well as some pictorial examples—notably, this answer (*https://oreil.ly/NR501*).

Although these command-line tools are old (for example, Unix development started in the late 1960s), people still use them because of their power. For example, deleting thousands of files would likely take many clicks with a mouse but only one line of code at the command line.

When starting out, command lines can be confusing, just like starting with Python or R. Likewise, using the command line is a fundamental skill, similar to running or coordination drills in football. The command line is also used with most of the advanced skills we list and will also enhance your understanding of languages such as R or Python. For example, understanding the command line will enhance your understanding of programming languages by making you think about file structures and how computer operating systems work. But which command line to use?

We suggest you consider two options. First, the *Bourne Again Shell* (shortened to *bash*, named after the Bourne shell that it supersedes, which was named after the shell's creator, Stephen Bourne) traditionally has been the default shell on Linux and macOS. This shell is also now available on Windows and is often the default for cloud computers (such as AWS, Microsoft Azure, and GCP) and high-performance supercomputers. Most likely, you will start with the bash shell.

A second option is Microsoft PowerShell. Historically, this was only for Windows, but is now available for other operating systems as well. PowerShell would be the best choice to learn if you also do a lot of information technology work in a corporate setting. The tools in PowerShell would be able to help you automate parts of your job such as security updates and software installs.

If you have macOS or Linux, you already have a terminal with a bash or bash-clone terminal (macOS switched to using the Zsh shell language because of copyright issues, but Zsh and bash are interchangeable for many situations, including our basic examples). Simply open the terminal app on your computer and follow along. If you use Windows, we suggest downloading Git for Windows (*https://gitforwindows.org*), which comes with a lightweight bash shell. For Windows users who discover the utility of bash, you may eventually want to move to using Windows Subsystem for Linux (WSL). This program gives you a powerful, complete version of Linux on your Windows computer.

Bash Example

A terminal interface forces you to think about file structure on your computer. When you open your terminal, type **pwd** to print (on the screen) the current working directory. For example, on Richard's Linux computer running Pop!_OS (a flavor of Linux), this looks like the following:

```
(base) raerickson@pop-os:~$ pwd
/home/raerickson
```

Here, *home/raerickson* is the current working directory. To see the files in the working directory, type the list command, **ls** (we also think about **ls** as being short for *list stuff* as a way to remember the command):

```
raerickson@pop-os:~$  ls
Desktop                 Games           Public
Documents               R               Untitled.ipynb
Downloads               miniconda3      Videos
Firefox_wallpaper.png   Music           Templates
Pictures                test.py
```

You can see all directories and files in Richard's user directory. Filepaths are also important. The three basic paths to know are your current directory, your computer's home directory, and up a level:

- ./ is the current directory.
- ~/ is your computer's default home directory.
- ../ is the previous directory.

For example, say your current directory is *home/raerickson*. With this example, your directory structure would look like this:

- ../ would be the *home* directory.
- ./ would be the *raerickson* directory.
- / would be the lowest level in your computer.
- ~/ would be the default home directory, which is *home/raerickson* on Richard's computer.

For practical purposes, *directory* and *folder* are the same term and you can use either with the examples in this book.

You can use the change directory command, cd, to change your current working directory. For example, to get to the home directory, you could type this:

```
cd ../
```

Or you could type this:

```
cd  /home/
```

The first option uses a *relative* path. The second option uses an *absolute* path. In general, relative paths are better than absolute, especially for languages like Python and R, when other people might be reusing code across multiple machines.

You can also use the command line to move files and directories. For example, to copy *test.py*, you need to make sure you are in the same directory as the file. To do this, use **cd** to navigate to the directory with *test.py*. Type **ls** to make sure you can see the file. Then use **cp** (the copy function) to copy the file to *Documents*.

```
cp test.py ./Documents
```

You can also use cp with different filepaths. For example, let's say you're in *Documents* and want to move *test.py* to *python_code*. You could use the filepaths with cp:

```
cp  ../test.py ./python_code
```

In this example, you are currently in */home/raerickson/Documents*. You can take the file from */home/raerickson/test.py/* by moving *../test.py* to the directory */home/raerickson/Documents/python_code* using *./python_code*.

You can also copy directories. To do this, use the recursive option (or, in Linux parlance, *flag*) -r with the copy command. For example, to copy *python_code*, you would use cp ./python_code new_location. A move function also exists, which does not leave the original object behind. The move command is mv.

 Command-line file deletions do not go to a recycling or trash directory on your computer. Deletions are permanent.

Lastly, you can remove directories and files by using the terminal. We recommend you be very careful. To delete, or remove, files, use rm *file_name*, where *file_name* is the file to delete. To delete a directory, use rm -r *directory* where *directory* is the directory you want to remove. To help you get started, Table 9-2 contains common bash commands we use on a regular basis.

Table 9-2. Common bash commands

Command	Name and description
pwd	Print working directory, to show your location
cd	Change directory, to change your location on your computer
cp	Copy a file
cp -r	Copy a directory
mv	Move a file
mv -r	Move a directory
rm	Remove a file
rm -r	Remove a directory

Suggested Readings for bash

The bash shell is not only a way to interact with your computer, but it also comes with its own programming language. We generally only touch the surface of its tools in our daily work, but some people extensively program in the language. Here are some bash resources to learn more:

- Software Carpentry (*https://software-carpentry.org*) offers free tutorials on the Unix Shell (*https://swcarpentry.github.io/shell-novice*). More generally, we recommend this site as a general resource for many of the topics covered in this book.

- *Learning the bash Shell*, 3rd edition, by Cameron Newham and Bill Rosenblatt (O'Reilly, 2005) provides introductions and coverage of the advanced tools and features of the language.

- *Data Science at the Command Line: Obtain, Scrub, Explore, and Model Data with Unix Power Tools*, 2nd edition by Jeroen Janssens (O'Reilly, 2021) shows how to use helpful command-line tools.

- We suggest your browse several books and find the one that meets your needs. Richard learned from *Unix Shells by Example* 4th edition, by Ellie Quigley (Pearson, 2004).

- Online vendors offer courses on bash. We do not have any recommendations.

Version Control

When working on code on a regular basis, we face problems such as *how do we keep track of changes?* or *how do we share code?* The solution to this problem is *version control software*. Historically, several programs existed. These programs emerged from the need to collaborate and see what others did as well as to keep track of your own changes to code. Currently, Git is the major version control program; around the time of publication, it has a high market share, ranging from 70% (*https://oreil.ly/6qO8r*) to 90% (*https://oreil.ly/nsZAX*).

Git emerged because Linus Torvalds faced that problem with the operating system he created, Linux. He needed a lightweight, efficient program to track changes from an army of volunteer programmers around the word. Existing programs used too much memory because they kept multiple versions of each file. Instead, he created a program that tracked only changes across files. He called this program *Git*.

> *Fun fact:* Linus Torvalds has, half-jokingly, claimed to name his two software programs after himself (*https://oreil.ly/UUkZk*). *Linux* is a recursive acronym— Linux is not Unix (Linux)—but it is also close to his own first name. *Git* is British English slang for an arrogant person or jerk. Torvalds, by his own admission, can be difficult to work with. As an example, searching for images of Torvalds will show him responding to a reporter's question with an obscene gesture.

Git

Git, at its heart, is an open source program that allows anybody to track changes to code. People can use Git on their own computer to track their own changes. We will start with some basic concepts of Git here. First, you need to obtain Git.

- For Windows users, we like Git for Windows (*https://www.gitforwindows.org*).

- For macOS users, we encourage you to make sure you have Terminal installed. If you install Xcode, Git will be included, but this will be an older version. Instead, we encourage you to upgrade Git from the Git project home page (*https://git-scm.com*).

- For Linux users, we encourage you to upgrade the Git that comes with your OS to be safe.

- For people wanting a GUI on Windows or macOS systems, we suggest you check out GitHub Desktop (*https://oreil.ly/Ghub*). The Git project page (*https://oreil.ly/bfYbS*) lists many other clients, including GUIs for Linux as well as Windows and macOS.

> Command-line Git is more powerful than any GUI but is also more difficult. We show the concepts using the command line, but we encourage you to use a GUI. Two good options include GitHub's GUI and the default Git GUI that comes with Git.

After obtaining Git, you need to tell Git where to keep track of your code:

1. Open a terminal.

2. Change your working directory to your project directory via **cd path/ to_my_code/**.

3. In one line, type **git init** then press Enter/Return. The git command tells the terminal to use the Git program, and init tells the Git program to use the init command.

4. Tell Git what code to track. You can do this for individual files via **git add filename** or for all files with **git add .** (The period is a shortcut for all files and directories in the current directory).

5. Commit your changes to the code with **git commit -m "initial commit"**. With this command, git tells the terminal which program to use. The commit command tells git to commit. The *flag* -m tells commit to accept the message in quotes, *"my changes"*. With future edits, you will want to use descriptive terms here.

Be careful with which files you track. Seldomly will you want to track data files (such as *.csv* files) or output files such as images or tables. Be extra careful if posting code to public repositories such as GitHub. You can use *.gitignore* files to block tracking for all file types via commands such as ***.csv** to block tracking CSV files.

Now, you may edit your code. Let's say you edit the file *my_code.R* and then change it. Type **git status** to see that this file has been changed. You may add the changes to the file by typing **git add my_code.R**. Then you need to commit the changes with **git commit -m "example changes"**.

The learning curve for Git pays itself off if you accidentally delete one or more important files. Rather than losing days, weeks, months, or longer of work, you lose only the amount of time it takes you to search for undoing a delete with Git. *Trust us; we know from experience and the school of hard knocks.*

GitHub and GitLab

After you become comfortable with Git (at least comfortable enough to start sharing your code), you will want to back up and share code. When Richard was in grad school, around 2007, he had to use a terminal to remotely log on to his advisor's computer and use Git to obtain code for his PhD project. He had to do this because easy-to-use commercial solutions (like GitHub) did not yet exist for sharing code. Luckily, commercial services now host Git repositories.

The largest provider is GitHub (*https://github.com*). This company and service are now owned by Microsoft. Its business model allows hosting for free but charges for business users and extra features for users. The second largest provider is GitLab (*https://gitlab.com*). It has a similar business model but is also more developer focused. GitLab also includes an option for free self-hosting using its open source software. For example, O'Reilly Media and one of our employers both self-host their own GitLab repositories.

Regardless of which commercial platform you use, all use the same underlying Git technology and command-line tools. Even though the providers offer different websites, GUI tools, and bells and whistles, the underlying Git program is the same. Our default go-to is GitHub, but we know that some people prefer to avoid Microsoft and use GitLab. Another choice is Bitbucket (*https://bitbucket.org*), but we are less familiar with this platform.

The purpose of a remote repository is that your code is backed up there and other people can also access it. If you want, you can share your code with others. For open source software, people can report bugs as well as contribute new features and bug fixes for your software. We also like GitHub's and GitLab's online GUIs because they allow people to see who has updated and changed code. Another feature we like about these web pages is that they will render Jupyter Notebook and Markdown files as static web pages.

GitHub Web Pages and Résumés

A fun way to learn about Git and GitHub is to build a résumé. A search for GitHub résumés should help you find online tutorials (we do not include links because these pages are constantly changing). A Git-based résumé allows you to show your skills and create a marketable product demonstrating your skills. You can also use this to show off your football products, whether for fun or as part of job hunting. For example, we will have our interns create these pages as a way to document what they have learned while also learning Git better. Former intern John Oliver has an example résumé at *https://oreil.ly/JOliv*.

Suggested Reading for Git

Many resources can help you learn about Git, depending on your learning style. Some resources we have found to be useful include the following:

- Git tutorials are offered on the Git project home page (*https://git-scm.com*), including videos that provide an overview of the Git technology for people with no background.
- Software Carpentry offers Git tutorials (*https://swcarpentry.github.io/git-novice*).

- Training materials on GitHub provide resources. We do not provide a direct link because these change over time.

- Git books from O'Reilly such as *Version Control with Git*, 3rd edition, by Prem Kumar Ponuthorai and Jon Loeliger (2022) can provide resources all in one location.

Style Guides and Linting

When we write, we use different styles. A text to our partner that we're at the store might be *At store, see u.* followed by the response *k. plz buy milk*. A report to our boss would be very different, and a report to an external client even more formal. Coding can also have different styles. Style guides exist to create consistent code. However, programmers are a creative and pragmatic bunch and have created tools to help themselves follow styles. Broadly, these tools are called *linting*.

The term *linting* comes from removing lint from clothes, like lint-rolling a sweater to remove specks of debris.

Different standards exist for different languages. For Python, PEP 8 is probably the most common style guide, although other style guides exist. For R, the Tidyverse/Google style guides are probably the most common style.

Open source projects often split and then rejoin, ultimately becoming woven together. R style guides are no exception. Google first created an R Style Guide. Then the Tidyverse Style Guide based itself upon Google's R Style Guide. But then, Google adapted the Tidyverse Style Guide for R with its own modifications. This interwoven history is described on the Tidyverse Style Guide page (*https://style.tidyverse.org*) and Google's R Style Guide page (*https://oreil.ly/RGuide*).

To learn more about the styles, please visit the PEP 8 Style Guide (*https://oreil.ly/PEP8*), Tidyverse style home pages (*https://style.tidyverse.org*), or Google's style guides for many languages (*https://oreil.ly/YLnE6*).

Google's style guides are hosted on GitHub using a Markdown language that is undoubtedly tracked using Git.

Table 9-1 lists some example linting programs. We also encourage you to look at the documentation for your code editor. These often will include add-ons (or plug-ins) that allow you to lint your code as you write.

Packages

We often end up writing custom functions while programming in Python or R. We need ways to easily reuse these functions and share them with others. We do this by placing the functions in packages. For example, Richard has created Bayesian models used for fisheries analysis that use the Stan language as called through R. He has released these models as an R package, fishStan (*https://oreil.ly/iJagB*). The outputs from these models are then used in a fisheries model, which has been released as a Python package (*https://oreil.ly/XCwTS*).

With a package, we keep all our functions in the same place. Not only does this allow for reuse, but it also allows us to fix one bug and not hunt down multiple versions of the same file. We can also include tests to make sure our functions work as expected, even after updating or changing functions. Thus, packages allow us to create reusable and easy-to-maintain code.

We can use packages to share code with others. Probably the most common way to release packages is on GitHub repos. Because of the low barrier to entry, anyone can release packages. Python also has multiple package managers including pip and conda-forge, where people can submit packages. Likewise, R currently has one major package manager (and historically, had more): the Comprehensive R Archive Network (CRAN). These repositories have different levels of quality standards prior to submission, and thus some gatekeeping occurs compared to a direct release on sites like GitHub.

Suggested Readings for Packages

- To learn about R packages, we used Hadley Wickham and Jennifer Bryan's online book (*https://r-pkgs.org*) that is also available as a dead-tree version from O'Reilly.

- To learn about Python packages, we used the Official Python tutorial, "Packaging Python Projects." (*https://oreil.ly/ySvav*)

Computer Environments

Imagine you run an R or Python package, but then the package does not run next session. Eventually, hours later, you figure out that a package was updated by its owner and now you need to update your code (yes, similar situations have occurred to us). One method to prevent this problem is to keep track of your computer's environment. Likewise, computer users might have problems working with others.

For example, Eric wrote this book on a Windows computer, whereas Richard used a Linux computer.

A computer's environment is the computer's collection of software and hardware. For example, you might be using a 2022 Dell XPS 13-inch laptop for your hardware. Your software environment might include your operating system (OS), such as Windows 11 release 22H2 (10.0.22621.1105), as well as the versions of R, Python, and their packages, such as R 4.1.3 with `ggplot2` version 3.4.0. In general, most people are concerned about the programs for the computing environment. When an environment does not match across users (for example, Richard and Eric) or across time (for example, Eric's computer in 2023 compared to 2021), programs will sometimes not run, such as the problem shown in Figure 9-4.

We cannot do justice for virtual environments like Conda in this book. However, many programmers and data scientists would argue their use helps differentiate experienced professionals from amateurs.

Tools like Conda (*https://oreil.ly/Conda*) let you lock down your computer's environment and share the specific programs used. Tools like Docker (*https://www.docker.com*) go a step further and control not only the environment but also the operating systems. Both of these programs work best when the user understands the terminal.

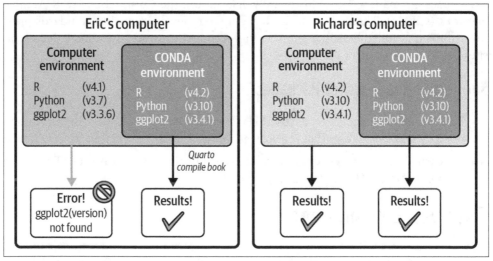

Figure 9-4. Example of computer environments and how versions may vary across users and machines

Interactives and Report Tools to Share Data

Most people do not code. However, many people want access to data, and, hopefully, you want to share code. Tools for sharing data and models include interactive applications, or *interactives*. These allow people to interact with your code and results. For small projects, such as the ones some users may want to share after completing this book, programs like Posit's Shiny (*https://shiny.posit.co*) or web-hosted Jupyter notebooks with widgets (*https://jupyter.org/widgets*) may meet your needs. People working in the data science industry, like Eric, will also use these tools to prototype models before handing the proof-of-concept tool over to a team of computer scientists to create a production-grade product.

Interactives work great for dynamic tools to see data. Other times, you may want or need to write reports. Markdown-based tools let you merge code, data, figures, and text all into one. For example, Eric writes reports to clients in R Markdown (*https://rmarkdown.rstudio.com*), Richard writes software documentation in Jupyter Notebook (*https://jupyter.org*), Richard writes scientific papers in LaTeX (*https://www.latex-project.org*), and this book was written in Quarto (*https://quarto.org*). If starting out, we suggest Quarto because the language expands upon R Markdown to also work with Python and other languages (R Markdown itself was created to be an easier-to-use version of LaTeX). Jupyter Notebook can also be helpful for reports and longer documents (for example, books have been written using Jupyter Notebook) but tend to work better for dynamic applications like interactives.

Artificial Intelligence Tools

Currently, tools exist that help people code using artificial intelligence (AI) or similar-type tools. For example, many code editors have autocompletion tools. These tools, at their core, are functionally AI. During the writing of this book, new AI tools have emerged that hold great potential to assist people coding. For example, ChatGPT can be used to generate code based on user input prompts. Likewise, programs such as GitHub Copilot help people code based on input prompts, and Google launched its own competing program, Codey.

However, AI tools are still new, and challenges exist with their use. For example, the tools will produce factual errors (*https://oreil.ly/nJOpQ*) and well-documented biases (*https://oreil.ly/K1IlO*). Besides well documented factual errors and biases, the programs consume user data. Although this helps create a better program through feedback, people can accidentally release data they did not intend to release. For example, Samsung staff accidentally released semiconductor software and proprietary data to ChatGPT (*https://oreil.ly/-rgCg*). Likewise, the Copilot for Business Privacy Statement (*https://oreil.ly/Nflm7*) notes that "it collects data to provide the service, some of which is then saved for further analysis and product improvements."

Do not upload data and code to AI services unless you understand how the services may use and store your data and code.

We predict that AI-based coding tools will greatly enhance coding but also require skilled operators. For example, spellcheckers and grammar checkers did not remove the need for editors. They simply reduced one part of editors' jobs.

Conclusion

American football is the most popular sport in the US and one of the most popular sports in the world. Millions of fans travel countless miles to see their favorite teams every year, and more than a billion dollars in television deals are signed every time a contract is ready to be renewed. Football is a great vehicle for all sorts of involvement, whether that be entertainment, leisure, pride, or investment. Now, hopefully, it is also a vehicle for math.

Throughout this book, we've laid out the various ways in which a mathematically inclined person could better understand the game through statistical and computational tools learned in many undergraduate programs throughout the world. These very same approaches have helped us as an industry push the conversation forward into new terrain, where analytically driven approaches are creating new problems for us to solve, which will no doubt create additional problems for you, the reader of this book, to solve in the future.

The last decade of football analytics has seen us move the conversation toward the "signal and the noise" framework popularized by Nate Silver. For example, Eric and his former coworker George Chahrouri asked the question "if we want to predict the future, should we put more stock in how a quarterback plays under pressure or from a clean pocket?" in a PFF article (*https://oreil.ly/JnD78*).

We've also seen a dramatic shift in the value of players by position, which has largely been in line with the work of analytical firms like PFF, which help people construct rosters by valuing positions (*https://oreil.ly/fkg1U*). Likewise, websites like *https://rbsdm.com* have allowed fans, analysts, and scribes to contextualize the game they love and/or cover using data.

On a similar note, the legalization of sports betting across much of the US has increased the need to be able to tease out the meaningful from the misleading. Even the NFL Draft, at one point an afterthought in the football calendar, has become a high-stakes poker game and, as such, has attracted the best minds in the game working on making player selection, and asset allocation, as efficient as possible.

The future is bright for football analytics as well. With the recent proliferation of player-tracking data, the insights from this book should serve as a jumping-off point in a field with an ever-growing set of problems that should make the game more enjoyable. After all, almost every analytical advancement in the game (more passing, more fourth-down attempts) has made the game more entertaining. We predict that trend will continue.

Furthermore, sports analytics in general, and football analytics specifically, has opened doors for many more people to participate in sports and be actively engaged than in previous generations. For example, Eric's internship program, as of May of 2023, has sent four people into NFL front offices, with hopefully many more to come. By expanding who can participate, as well as add value to the game, football now has the opportunity to become a lot more compelling for future generations.

Hopefully, our book has increased your interest in football and football analytics. For someone who just wants to dabble in this up-and-coming field, perhaps it contains everything you need. For those who want to gain an edge in fantasy football or in your office pool, you can update our examples to get the current year's data. For people seeking to dive deeper, the references in each chapter should provide a jumping-off point for future inquiry.

Lastly, here are some websites to look at for more football information:

- Eric's current employer, SumerSports: *https://sumersports.com/*
- Eric's former employer, PFF: *https://www.pff.com*
- A website focusing on advanced NFL stats, Football Outsiders: *https://www.footballoutsiders.com*
- Ben Baldwin's page that links to many other great resources: *https://rbsdm.com*

Happy coding as you dive deeper into football data!

Python and R Basics

This appendix provides details on the basics of Python and R. For readers unfamiliar with either language, we recommend you spend an hour or two working through basic tutorials such as those listed on the project's home pages listed in "Local Installation" on page 262.

 In both Python and R, you may access built-in help files for a function by typing **help("function")**. For example, typing help("+") loads a help file for the addition operator. In R, you can often also use the ? shortcut (for example ?"+"), but you need to use quotes for some operators such as + or -.

Obtaining Python and R

Two options exist for running Python or R. You may install the program locally on a computer or may use a cloud-hosted version. From the arrival of the *personal computer* in the late-1980s up until the 2010s, almost all computing was local. People downloaded and installed software to their local computer unless their employer hosted (and paid for) cloud access or local servers. Most people still use this approach and use local computing rather than the cloud. More recently, consumer-scale cloud tools have become readily available. Trade-offs exist for using both approaches:

- Benefits of using a personal computer:
    - You have known, up-front, one-time costs for hardware.
    - You control all hardware and software (including storage devices).
    - You can install anything you want.

— Your program will run without internet connections, such as during internet outages, in remote locations, or while flying.

- Benefits of using consumer-scale cloud-based computing:

   — You pay for only the computing resources you need.

   — Many free consumer services exist.

   — You do not need expensive hardware; for example, you could use a Chromebook or even tablets such as an iPad.

   — Cloud-based computing is cheaper to upscale when you need more resources.

   — Depending on platform and support levels, you're not responsible for installing programs.

Local Installation

Directions for installing Python and R appear on each project's respective pages:

- Python (*https://www.python.org*)
- R Project (*https://www.r-project.org*)

We do not include directions here because the programs may change and depend on your specific operating systems.

Cloud-Based Options

Many vendors offer cloud-based Python and R. Vendors in 2023 include the following:

- Google Colab (*https://colab.research.google.com*)
- Posit Cloud (*https://posit.cloud*)
- DataCamp Workspaces (*https://www.datacamp.com/workspace*)
- O'Reilly learning platform (*https://learning.oreilly.com/home*)

Scripts

Script files (or *scripts*, for short) allow you to save your work for future applications of your code. Although our examples show typing directly in the terminal, typing in the terminal is not effective or easy. These files typically have an ending specific for the language. Python files end in *.py*, and R files end in *.R*.

During the course of this book, you will be using Python and R interactively. However, some people also run these languages as batch files. A *batch file* simply tells the computer to run an entire file and then spits out the outputs from the file. An example batch file might calculate summary statistics that are run weekly during the NFL season by a company like PFF and then placed into client reports.

In addition to batch files, some people have large amounts of text needed to describe their code or code outputs. Other formats often work better than script files for these applications. For example, Jupyter Notebook allows you to create, embed, and use runnable code with easy-to-read documents. The *Notebooks* in the name "Jupyter Notebook" suggests a similarity to a scientist's laboratory or field notebook, where text and observations become intermingled.

Likewise, Markdown-based files (such as Quarto or the older R Markdown files) allow you to embed static code and code outputs into documents. We use both for our work. When we want other people to be readily able to interactively run our code, we use Jupyter Notebook. When we want to create static reports such as project files for NFL teams or scientific articles, we use R Markdown-based workflows.

Early drafts of this book were written in Jupyter Notebook because we were not aware of O'Reilly's support for Quarto-based workflows. Later drafts were written with Quarto because we (as authors) know this language better. Additionally, Quarto worked better with both Python and R in the same document. Hence we, as authors, adapted our tool choice based on our publisher's systems, which illustrates the importance of flexibility in your toolbox.

 Many programs, such as Microsoft Word, use "smart quotes" and other text formatting. Because of this, be wary of any code copied over from files that are not plain-text files such as PDFs, Word documents, or web pages.

Packages in Python and R

Both Python and R succeed and thrive as languages because they can be extended via packages. Python has multiple package-management systems, which can be confusing for novice and advanced users. Novice users often don't know which system to use, whereas advanced users run into conflicting versions of packages across systems.

The simplest method to install Python packages is probably from the terminal outside of Python, using the `pip` command. For example, you may install `seaborn` by using `pip install seaborn`. R currently has only one major repository of interest to readers of this book, the Comprehensive R Archive Network (CRAN); a second repository, Bioconductor, mainly supports bioinformatics projects. An R package may be installed inside R by using the `install.packages()` function. For example, you may install `ggplot2` with `install.packages("ggplot2")`.

You may also, at some point, find yourself needing to install packages from sites like GitHub because the package is either not on CRAN or you need the development version to fix a bug that is blocking you from using your code. Although these installation methods are beyond the scope of this book, you can easily find them by using a search engine. Furthermore, environment-management software exists that allows you to lock down package versions. We discussed these in Chapter 9.

nflfastR and nfl_data_py Tips

This section provides our specific observations about the `nflfastR` and `nfl_data_py` packages. First, we (the authors) love these packages. They are great free resources of data.

Update the `nflfastR` and `nfl_data_py` packages on a regular basis throughout the season as more data is added from new games. That will give you access to the newest data.

You may have noticed that you have to do a fair amount of wrangling with data. That's not a downfall of these packages. Instead, this shows the flexibility and depth of these data sources. We also have you clean up the datasets, rather than provide you with clean datasets, because we want you to learn how to clean up your data. Also, these datasets are updated nightly during the football season, so, by showing you how to use these packages, we give you the tools to obtain your own data.

To learn more about both packages, we encourage you to dive into their details. Once you've looked over the basic functions and structure of the packages, you may even want to look at the code and "peek under the hood." Both `nflfastR` (*https://oreil.ly/B6qiV*) and `nfl_data_py` (*https://oreil.ly/ss6d_*) are on GitHub. This site allows you to report issues and bugs, and to suggest changes—ideally by providing your own code!

Lastly, we encourage you to give back to the open source community that supplies these tools. Although not everyone can contribute code, helping with other tasks like documentation may be more accessible.

Integrated Development Environments

You can use powerful code-editing tools called *integrated development environments* (*IDEs*). Much like football fans fight over who is the best quarterback of all time, programmers often argue over which IDEs are best. Although powerful (for example, IDEs often include syntax checkers similar to a spellchecker and autocompletion tools), IDEs can have downsides.

Some IDEs are complex, which can be great for expert users, but overwhelming for beginners and casual users. For example, the Emacs text editor has been jokingly described as an operating system with a good text editor or two built into it. Others, such as the web comic "xkcd," poke fun at IDEs such as emacs as well (*https:// xkcd.com/378*). Likewise, some professional programmers feel that the shortcuts built into some IDEs limit or constrain understanding of languages because they do not require the programmer to have as deep of an understanding of the language they are working in.

However, for most users, especially casual users, the benefits of IDEs far outweigh the downsides. If you already use another IDE for a different language at work or elsewhere, that IDE likely works with Python and possibly R as well.

When writing this book, we used different editors at different times and for different file types. Editors we used included RStudio Desktop, JupyterLab, Visual Studio Code, and Emacs. Many good IDEs exist, including many we do not list. Some options include the following:

- Microsoft Visual Studio Code (*https://code.visualstudio.com*)
- Posit RStudio Desktop (*https://posit.co*)
- JetBrain PyCharm (*https://www.jetbrains.com/pycharm*)
- Project Jupyter JupyterLab (*https://jupyter.org*)
- GNU Project Emacs (*https://www.gnu.org/software/emacs*)

Basic Python Data Types

As a crash course in Python, various types of objects exist within the language. The most basic types, at least for this book, include *integers*, *floating-point numbers*, *logical*, *strings*, *lists*, and *dictionaries*. In general, Python takes care of thinking about these data types for you and will usually change the type of number for you.

Integers (or *ints* for short) are whole numbers like 1, 2, and 3. Sometimes, you need integers to index objects (for example, taking the second element, for a vector x by using x[1]; recall that Python starts counting with 0). For example, you can create an integer x in Python:

```
## Python
x = 1
x.__class__
```

Resulting in:

```
<class 'int'>
```

Floating-point numbers (or *floats*, for short) are decimal numbers that computers keep track of to a finite number of digits, such as `float16`. Python will turn ints into floats when needed, as shown here:

```Python
## Python
y = x/2
y.__class__
```

Resulting in:

```
<class 'float'>
```

Computers cannot remember all digits for a number (for example, `float16` keeps track of only 16 digits), and computer numbers are not mathematical numbers. Consider the parlor trick Richard's math co-advisor taught him in grad school:

```Python
## Python
1 + 1 + 1 == 3
```

Resulting in:

```
True
```

But, this does not always work:

```Python
## Python
0.1 + 0.1 + 0.1 == 0.3
```

Resulting in:

```
False
```

Why does this not work? The computer chip has a rounding error that occurs when calculating numbers. Hence, the computer "fails" at mathematical addition.

Python also has `True` and `False` logical operators that are called *Boolean objects*, or *bools*. The previous examples showed how bools result from logical operators.

Letters are stored as *strings* (or *str*, for short). Numbers stored as strings do not behave like numbers. For example, look at this:

```Python
## Python
a = "my two"
c = 2
a * c
```

Resulting in:

```
'my twomy two'
```

Python has groups of values. Lists may be created using `list()` or `[]`:

```Python
## Python
my_list = [1, 2, 3]
my_list_2 = list(("a", "b", "c"))
```

Dictionaries may also store values with a *key* and are created with dict() or {}. We use the {} almost exclusively but show the other method for completeness. For example:

```Python
## Python
my_dict = dict([('fred', 2), ('sally', 5), ('joe', 3)])
my_dict_2 = {"a": 0, "b": 1}
```

Additionally, the numpy package allows data to be stored in arrays, and the pandas package allows data to be stored in dataframes. Arrays must be the same data type, but dataframes may have mixed columns (such as one column of numbers and a second column of names).

Basic R Data Types

As a crash course in R, multiple types of objects exist within the language, as in Python. The most basic types, as least for this book, include *integers, numeric floating-point numbers, logical, characters,* and *lists*. In general, R takes care of thinking about these structures for you and will usually change the type for you. R uses slightly different names than Python; hence this section is similar to "Basic Python Data Types" on page 265, but slightly different.

Integers are whole numbers like 1, 2, and 3. Sometimes, you need integers to index objects—for example, taking the second element for a vector x by using x[2]; recall that R starts counting with 1. For example, you can create an integer x in R by using L:

```R
## R
x <- 1L
class(x)
```

Resulting in:

```
[1] "integer"
```

Numerical floating-point numbers (or *numeric,* for short) are decimal numbers that computers keep track of to a finite number of digits. R will turn integers into numerics when needed, as shown here:

```R
## R
y <- x / 2
class(y)
```

Resulting in:

```
[1] "numeric"
```

Computers cannot remember all digits for a number (for example, float16 keeps track of only 16 digits), and computer numbers are not mathematical numbers. Consider the parlor trick Richard's math co-advisor taught him in grad school:

```
## R
1 + 1 + 1 == 3
```

Resulting in:

```
[1] TRUE
```

But this does not always work:

```
## R
0.1 + 0.1 + 0.1 == 0.3
```

Resulting in:

```
[1] FALSE
```

This is due to a rounding error that occurs on the computer chip calculating the answer.

R also has TRUE and FALSE logical operators that are called *logical* for short. The previous examples showed how logical outputs result from logical operators.

Letters are stored as *characters* in R. These do not work with numeric operators. For example, look at the following:

```
## R
a <- "my two"
c <- 2
a * c
```

R has groups of values called *lists* or *vectors*. Lists may be created by using a *combine* or *concatenate* function. For example:

```
## R
my_list <- c(1, 2, 3)
```

Additionally, base R contains matrices to store numbers and dataframes to store mixed columns (such as one column of numbers and a second column of names). We use tibbles from the tidyverse in this book as an upgraded version of dataframes.

Summary Statistics and Data Wrangling: Passing the Ball

This appendix contains materials to help you understand some basic statistics. If the topics are new to you, we encourage you to read this material after Chapter 1 or Chapter 2 and before you dive too far into the book.

In Chapter 2, you looked at quarterback performance at different pass depths in an effort to understand which aspect of play was fundamental to performance and which aspect was noisier, possibly leading you astray as you aimed to make predictions about future performance. You were lucky enough to have the data more or less in ready-made form for you to perform this analysis. You did have to create your own variable for analysis, but such *data wrangling* was minimal at best.

Sports analytics generally, and football analytics specifically, are still in their early stages of development. As such, datasets may not always be the cleanest, or tidy. *Tidy* datasets are usually in a table form that computers can easily read and humans can easily understand. Furthermore, data analysis in any field (and football analytics is no different) often requires datasets that were created for different purposes. This is where data wrangling can come in handy. Because so many people have had to clean up messy data, many terms exist in this field. Some synonyms for *data wrangling* include *data cleaning*, *data manipulating*, *data mutating*, *shaping*, *tidying*, and *munging*. More specifically, these terms describe the process of using a programming language such as Python or R to update datasets to meet your needs.

Tidy data has two definitions. Broadly, it refers to clean data. More formally, Hadley Wickham defines the term in a paper titled "Tidy Data." (*https://oreil.ly/tidy*) Wickham defines *tidy data* as having a specific structure: "(1) Each variable forms a column, (2) each observation forms a row, and (3) each type of observational unit forms a table."

During the course of our careers, we have found that data wrangling takes the most time for our projects. For example, one of our bosses once pinged us on Google chat because he was having trouble fitting a new model. His problem turned out to not be the model, but rather data formatting. Figuring out how to format the data to work with the model took about 30 minutes. However, running the new model took only about 30 seconds in R after we figured out the data structure issue he was having.

 Computer tools are ever changing, and data wrangling is no exception. During the course of his career, Richard has had to learn four tools for data wrangling: base R (around 2007), `data.table` in R (around 2012), the `tidyverse` in R (around 2015), and `pandas` in Python (around 2020). Hence, you will also likely need to update your skill set for the tools taught in this book. However, the fundamentals never change as long as you understand the basics.

Programming languages like Python or R are our most-effective tools for making our data to usable. The languages allow scripting, thereby letting us track our changes and see what we did, including whether we introduced any errors into our data.

Many people like to use spreadsheet programs such as Microsoft Excel or Google Sheets for data manipulation. Unfortunately, these programs do not easily keep track of changes. Likewise, hand-editing data does not scale, so as the size of the problem becomes too large—such as when you are working with player *tracking data*, which has one record for every player, anywhere from 10 to 25 times per second per play—you will not be able to quickly and efficiently build a workflow that works in a spreadsheet environment. Thus, editing one or two files by hand is easy to do with Excel, but editing one or two thousand files by hand is not easy.

Conversely, programming languages, such as Python or R, readily scale. For example, if you have to format data after each week's games, Python or R could easily be used as part of a data pipeline, but spreadsheet-based data wrangling would be difficult to automate into a data pipeline.

A *data pipeline* is the flow of data as it moves from one location to another and undergoes changes such as formatting. For example, when Eric worked at PFF, a data pipeline might take weekly numbers, format the numbers, run a model, and then generate a report. In computer science, a *pipe* operator refers to passing the outputs from one function directly to another.

That being said, we understand that many people like to use tools they are familiar with. If you are switching over to Python or R from using programs like Excel, we encourage you to switch one step at a time. As an analogy, think about a cook licking the batter spoon to taste the dish. When cooking at home for your family, many people do this. But a chef at a restaurant would hopefully be fired for licking and reusing their spoon. Likewise, recreational data analysis can reasonably use a program like Excel to edit data. But professional data analysis requires the use of code to wrangle data.

We encourage you to start doing one step at a time in Python or R if you already use a program like Excel. For example, let's say you currently format your football data in Excel, plot the data in Excel, and then fit a linear regression model in Excel. Start by plotting your data in Python or R the next time you work with your data. Once you get the hang of that, start fitting your model in Python or R. Finally, switch to formatting data in Python or R. For help with this transition, *Advancing into Analytics: From Excel to Python and R* by George Mount (O'Reilly, 2021) provides resources.

Besides data wrangling, you will also learn about some basic statistics in this appendix. *Statistics* means different things to different people.

During our day jobs, we see four uses of the word. Commonly, people use the word to refer to raw or objective data. For example, the (x, y) coordinates of a targeted pass might be referred to as the *stats* for a play. Sometimes a *statistic* can be something that is estimated, like expected points added (EPA) per play, or completion percentage above expected by a quarterback or offense. More formally, *statistics* can refer to the systematic collection and analysis of data. For example, somebody might run statistical analysis as part of a science experiment or as a business analyst using data science. Finally, the corresponding field of study related to the collection and analysis of data is called *statistics*. For example, you might have taken a statistics course in high school or know somebody who works as a professional statistician.

This appendix focuses on the use of *statistic* as something that can be estimated (specifically, summary statistics), and we show you how to summarize data by using statistics. For example, rather than needing to read the play-by-play report for a game, can you get an understanding of what occurred by looking at the summary statistics from the game? In Eric's job, he generally doesn't have the time to watch every game even once, let alone multiple times, nor does he have the chance to manually pore through each game's play-by-play data. As such, he generally builds systems that can deliver *key performance indicators* (*KPIs*) that can help him see trends emerge in an efficient way. Summary statistics can also serve as the *features* for models. For example, if someone wants to bet on the number of touchdowns a quarterback will throw for in a certain game, his average number of touchdown passes thrown in a game is likely to be very helpful.

Basic Statistics

Although not as glamorous as plotting, basic summary statistics are often more important because they serve as a foundation for data analysis and many plots.

Averages

Perhaps the simplest statistic is the *average*, or *mean*, or for the mathematically minded, the *expectation* of a set of numbers. Commonly, when we talk about the *average* for a dataset, we are talking about the central tendency of the data, the value that "balances" the data. We show how to calculate these by hand in the next section.

We intentionally do not include code for this section, as one of Eric's professors at the University of Nebraska–Lincoln, Dave Logan, would say, "The details of most calculations should be done once in everyone's life, but not twice." We will show you how to use Python and R to calculate these later in this appendix.

To work through this exercise, let's explore passing plays again, as shown in Chapter 1. This time, you will look at a quarterback's `air_yards` and study its properties in an attempt to understand his and his team's approach to the passing game. We use data from a 2020 game between Richard's favorite team, the Green Bay Packers, and one of their division rivals, the Detroit Lions, and only then look at plays that have an `air_yards` reading by Detroit over the middle of the field. "Filtering and Selecting Columns" on page 278 shows how to obtain and filter this data. Looking at a small set of data will allow you to easily do hand calculations.

First, calculate a *mean* by hand. The air yards are 5, –1, 5, 8, 5, 6, 1, 0, 16, and 17. To calculate the mean, first *sum* (or add up) the numbers:

$$5 + -1 + 5 + 8 + 6 + 5 + 6 + 1 + 0 + 16 + 17 = 68$$

Next, divide by the total number of plays with air yards:

$$\frac{68}{11} = 6.2$$

This allows you to estimate the mean air yards to the middle of the field to be 6.2 yards for Detroit during its first game against Green Bay in 2020. Also, we rounded the output to be 6.2. We need to round because the decimal of the resulting mean does not end and we really need to know only the first digit after the decimal, since the data is in integers. More formally, this is known as the number of *significant digits* or figures.

 Significant digits are important when reporting results. Although formal rules exist, a rule of thumb that works most of the time is to simply report the number of digits that are useful.

Another way to estimate the "center" of a data set is to examine the median. The *median* is simply the middle number, or the value of the average individual (rather than the average value). To calculate the median, write the numbers in order from smallest to largest and then find the middle number, or the average of the two middle numbers if you have an even number of values.

The last common method to estimate an average number is to examine the mode. The *mode* is the most common value in the dataset. To calculate the mode, we need to create a table with counts and air yards, such as Table B-1.

Table B-1. Summary table of number of passes for each air yard.

air_yards	count
−1	1
0	1
1	1
5	3
6	2
8	1
16	1
17	1

With this example, 5 is the mode because three observations provide a reading of 5 air yards. Data can be multimodal (that is to say, have multiple modes). For example, if two outcomes have the same number of occurrences, a bimodal outcome occurred.

Each central tendency measure has its pluses and minuses. The mean of a set of numbers depends heavily on outliers (see "Boxplots" on page 35 for a formal definition of *outliers*). If 2022 NFL MVP Patrick Mahomes, who at the time of this writing makes about $45 million annually, walks over to a set of nine practice squad players (each making $207,000 annually as of 2022), the average salary of the group is about $4.5

million per player, which doesn't accurately represent anything about that group. The median and mode of this dataset doesn't change at all ($207,000) with the inclusion of Mahomes, as the middle player scoots over half of a spot, and the group with the highest reading is the practice squad salary. That being said, much of the theoretical mathematics that has been built works a lot better with the mean, and discarding data points because they are outside the confines of the middle of the data set is not great practice, either. Thus, as with everything, the answer to which one you should use is "it depends."

Finally, other kinds of means exist, but they do not appear in this book. For example, in sports betting (or financial investing in general), one will often care about the geometric mean of a dataset rather than the arithmetic mean (which we've computed previously). The *geometric mean* is simply computed by multiplying all the numbers in the dataset together and then taking the root corresponding to the number of elements in the dataset. The reason this is preferable in betting or financial markets is clear: numbers grow (or decline) exponentially in this case rather than additively.

Let's examine these three measures of central tendency for air yards for all pass locations for both teams from the first game between Green Bay and Detroit in 2020 in Figure B-1. This subset of the data is more interesting to examine but would have been harder to examine *by hand*.

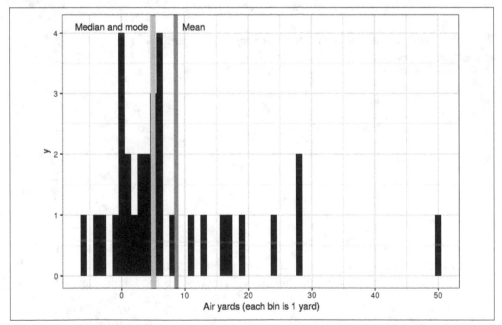

Figure B-1. Comparing different types of averages, also known as the central tendency of the data

---

First, notice that the line, labeled *mean* (blue online), is to the right of the median (red online). This means the data is skewed or has outliers to the right. Second, the median is the same as the mode in this example, so their lines overlap. Most people, including us, usually refer to the mean as the *average*.

So, what does this tell us about Detroit's passing in this game? First, the difference between mean and median shows that many pass plays are short, with the exception of a few long passes. This histogram shows us the "shape" of the data and the story it tells us about the game, which are probably better than simple summary statistics. This doesn't necessarily scale up with the size of the data, but it shows why it's always good to plot your data when you start and the reason we covered plotting as part of exploratory data analysis (EDA) in Chapter 2.

Variability and Distribution

The previous section dealt with central tendency. Many situations, however, require you to know about the *variability* in the data. The easiest way to examine the variability in the data is the *range*, which is the distance between the smallest (minimum, or min) and biggest number (maximum, or max) in the data. For data shown in Table B-1, the min is –1 and the max is 17, so the range is $17 - -1 = 17 + 1 = 18$.

Another method to examine the range and distribution of a dataset is to examine the *quantiles*. These focus on specific parts of the distribution, and their use can avoid the strength of severe outliers. The nth quantile is the data point where n% percent of the data lies underneath that data point. You've already seen one of these before, as the 50th quantile is the same thing as the median. "Boxplots" on page 35 covered quantiles (specifically within the context of boxplots). The 25th, 50th, and 75th quantiles are commonly referred to as *quartiles*, and the difference between the first quartile and the third is the *interquartile range (IQR)*.

Recall that boxplots show us where the middle 50% of the data occurs. Sometimes, other types of quantiles may be used as well. The benefit of quantiles are that they are estimated end points other than the central tenancy. For example, the mean or median allows you to examine how well average players do, but a quantile allows you to examine how well the best players do (for example, what does a player need in order to be better than 95% of other players?).

The most common method for examining the variability in a dataset is to look at the variance and its square root, the *standard deviation*. Broadly, the *variance* is the average squared deviation between the data points and the mean. The square is used so that the distance between each data point and the mean is counted as positive—so variability beneath and above the mean don't "cancel out." Using the Detroit Lions air yards example, you can do this calculation by hand with the numbers supplied in Table B-2. We include a mean column in case you are doing this calculation in a spreadsheet such as Excel and to also help you see where the numbers come from.

Table B-2. Calculating the difference and difference squared by hand using a spreadsheet

air_yards	mean	difference	difference squared
5	8.03125	−3.03125	9.18848
13	8.03125	4.96875	24.6885
3	8.03125	−5.03125	25.3135
6	8.03125	−2.03125	4.12598
6	8.03125	−2.03125	4.12598
−1	8.03125	−9.03125	81.5635
5	8.03125	−3.03125	9.18848
3	8.03125	−5.03125	25.3135
4	8.03125	−4.03125	16.251
28	8.03125	19.9688	398.751
28	8.03125	19.9688	398.751
11	8.03125	2.96875	8.81348
−6	8.03125	−14.0312	196.876
−4	8.03125	−12.0312	144.751
−3	8.03125	−11.0312	121.688
0	8.03125	−8.03125	64.501
8	8.03125	−0.03125	0.000976562
1	8.03125	−7.03125	49.4385
6	8.03125	−2.03125	4.12598
2	8.03125	−6.03125	36.376
5	8.03125	−3.03125	9.18848
6	8.03125	−2.03125	4.12598
19	8.03125	10.9688	120.313
0	8.03125	−8.03125	64.501
0	8.03125	−8.03125	64.501
1	8.03125	−7.03125	49.4385
4	8.03125	−4.03125	16.251
24	8.03125	15.9688	255.001
0	8.03125	−8.03125	64.501
16	8.03125	7.96875	63.501
17	8.03125	8.96875	80.4385
50	8.03125	41.9688	1761.38

After you create this table, you then take the sum of the difference squared column, which is 4,177. You divide that result by 76 because this is the number of observations minus 1. The reason to subtract 1 from the denominator has to do with the number of *degrees of freedom* at dataset has—the amount of data necessary to have a unique answer to a mathematical question.

Calculating $\frac{4,177}{(77-1)}$ gives the variance, 54.96. The units for variance with this example would be yards × yards, or yards². This does not relate that well to the central tendency of the data, so we take the square root to get the standard deviation: 7.41 yards, which is now in the units of the original data. All statistical software can calculate this value for you, and comparing variances and standard deviations across various datasets helps you compare the variability of multiple sources of data easily and in a way that scales up. Often, you will divide the standard deviation by the mean to get the *coefficient of variation*, which is a unit-less measure of variability that takes into account the size of data points. This might be important when comparing, say, passing yards per play to kickoff returns per play. One is on the order of 10, and the other is on the order of 20, and the variability understandably scales with that.

Uncertainty Around Estimates

When people give you predictions or summaries, a reasonable question is how much certainty exists around the prediction or summary? You can show uncertainty around the mean using the *standard error of the mean*, often abbreviated as *SEM*, or simply *SE*, for *standard error*. More informatively, you can estimate *confidence intervals (CIs)*. The most commonly used CI is 95% because of historical convention from statistics. The CI will contain the true or correct estimate 95% of the time if we repeat our observation process many, many times. If you accept this probability view of the world, you will know your CIs will include the mean 95% of the time, but you just won't know which 95% of the time.

The standard error is not the same thing as the standard deviation, but they are related. The former is trying to find the variability in the estimate of the population's mean, while the latter is trying to find the variability in the population itself. To compute the standard error from the standard deviation, you simply divide the standard deviation by the square root of the sample size. Thus, as the sample size n grows, your estimate for the population's mean becomes tighter, and the standard error decreases, while the variability in the population is fixed the whole time.

To calculate upper and lower bound of a confidence interval, use the *empirical rule* as a guide. The empirical says that, approximately 68% of the data lands within one standard deviation of the data, 95% within two standard deviations, and 99.7% within three. These values change with different distributions, and work best with a normal distribution (also known as a bell curve), but are pretty stable with respect to situation. The actual value for a 95% CI is 1.96.

Continuing with the previous example, calculate the standard error: $7.41 / \sqrt{77} = 0.84$. Thus, you can write the mean as $8.03 \pm 0.84(SE)$, or with the 95% CI as 8.03 (95% CI 6.38 to 9.68), which is calculated by $8.03 \pm 0.84 \times 1.96 = 1.64$.

> Always include uncertainty, such as a CI, around estimates such as mean when presenting to a technical audience. Presenting a *naked* mean is considered bad form because it does not allow the reader to see how much uncertainty exists around an estimate.

Based on statistical convention, you can compare 95% CIs to examine whether estimates differ. For example, the estimate of 8.03 (6.38 to 9.68 95% CI) differs from 0 because the 95% CI does not include 0. Thus, you can say that the air yards by the Detroit Lions in week 2 of 2020 differ from 0 when accounting for statistical uncertainty. If you were comparing two estimated means, you could compare both 95% CIs. If the CIs did not overlap, you can say the means are statistically different.

> People use 5% / 95% out of convention. Ronald L. Wasserstein et al. discuss this in a 2019 editorial in the *The American Statistician* (*https://oreil.ly/k_83m*). Their editorial presents many perspectives on alternative methods for statistical inference.

Chapter 3 and other chapters cover more about statistical inferences. Chapter 5 also covers more about methods for estimating variances and CIs. Now, enough about theory and hand calculations; let's see how to estimate these values in Python and R.

Filtering and Selecting Columns

To calculate summary statistics with Python and R, first load your data and the required R packages:

```
## R
library(tidyverse)
library(nflfastR)

# Load all data
pbp_r <- load_pbp(2020)
```

You can use similar code in Python:

```
## Python
import pandas as pd
import numpy as np
import nfl_data_py as nfl

# Load all data
pbp_py = nfl.import_pbp_data([2020])
```

Resulting in:

```
2020 done.
Downcasting floats.
```

After loading the data, select a subset of the data you want to use. Filtering or querying data is a fundamental skill for data science. At its core, filtering data uses logic statements. These statements can be really frustrating at times; never assume that you've done it correctly the first time. Richard remembers spending half a day in grad school stuck in the computer lab trying to filter out example air-quality data with R. Now, this task takes him about 30 seconds.

 Logic operators simply refer to computer code that compares a statement and provides a binary response. In Python, logical results are either True or False. In R, logical results are either TRUE or FALSE.

Python and R have different methods for filtering data. We've focused on the tools we use, but other useful approaches exist. For example, pandas dataframes have a .query() function that we like to use because it is more compact than .loc[]. However, some filtering requires .loc[] because .query() does not work in all situations. Likewise, the tidyverse in R has a filter() function. You can use these functions with logical operators.

In R, this can be done using the filter() function. Two true statements may be combined with an *and* (&) symbol. For example, select Green Bay (GB) as the home_team and Detroit (DET) as the away team, and then use home_team == 'GB' & away_team == 'DET' with the filter() function. Likewise, the select() function allows you to work with only the columns you need, creating a smaller and easier-to-use dataframe:

```
## R
# Filter out game data
gb_det_2020_r <-
    pbp_r |>
    filter(home_team == 'GB' & away_team == 'DET')

# select pass data
gb_det_2020_pass_r <-
    gb_det_2020_r |>
    select(posteam, yards_after_catch, air_yards,
            pass_location, qb_scramble)
```

Python uses a query() function. The input requires a set of quotes around the input, unlike R (for example, pandas uses "home_team == 'GB' & away_team == 'DET'"). In addition, pandas uses a list of column names to select specific columns of interest:

```
## Python
# Filter out game data
gb_det_2020_py = \
    pbp_py.query("home_team == 'GB' & away_team == 'DET'")

# Select pass some pass related columns
gb_det_2020_pass_py = \
    gb_det_2020_py[
        ["posteam", "yards_after_catch",
        "air_yards", "pass_location",
        "qb_scramble"]]
```

Calculating Summary Statistics with Python and R

With our dataset in hand, you can calculate the summary statistics introduced previously using Python and R. In Python, use describe() to see similar summaries that also include the median, count, minimum, and maximum values:

```
## Python
print(gb_det_2020_pass_py.describe())
```

Resulting in:

```
       yards_after_catch  air_yards  qb_scramble
count         38.000000  62.000000   181.000000
mean           6.263158   8.612903     0.016575
std            5.912352  10.938509     0.128025
min           -2.000000  -6.000000     0.000000
25%            2.250000   1.250000     0.000000
50%            4.000000   5.000000     0.000000
75%            9.000000  12.750000     0.000000
max           20.000000  50.000000     1.000000
```

In R, use the summary() function:

```
## R
summary(gb_det_2020_pass_r)
```

Resulting in:

```
   posteam            yards_after_catch     air_yards          pass_location
 Length:181         Min.   :-2.000       Min.   :-6.000      Length:181
 Class :character   1st Qu.: 2.250       1st Qu.: 1.250      Class :character
 Mode  :character   Median : 4.000       Median : 5.000      Mode  :character
                    Mean   : 6.263       Mean   : 8.613
                    3rd Qu.: 9.000       3rd Qu.:12.750
                    Max.   :20.000       Max.   :50.000
                    NA's   :143          NA's   :119

  qb_scramble
 Min.   :0.00000
 1st Qu.:0.00000
 Median :0.00000
 Mean   :0.01657
```

```
3rd Qu.:0.00000
Max.   :1.00000
```

One benefit of using summary() is that it shows the missing or NA values in R. This can help you see possible problems in the data. R also includes 1st Qu. and 3rd Qu., which are the first and third quartiles, which as stated previously are two of the special quantiles.

Both languages allow you to create customized summaries. For Python, use the .agg() function to aggregate the data frame. Use a dictionary inside Python to tell pandas which column to aggregate and what functions to use. Recall that Python defines dictionaries by using {"key" : [values]} notation for shorter notation or dict("key" : [values]) for a longer notation. In this case, the dictionary uses the column air_yards as the key and the aggregating functions as the list values:

```
## Python
print(gb_det_2020_pass_py.agg(
    {
        "air_yards": ["min", "max", "mean", "median",
                      "std", "var", "count"]
    }
))
```

Resulting in:

```
        air_yards
min      -6.000000
max      50.000000
mean      8.612903
median    5.000000
std      10.938509
var     119.650978
count    62.000000
```

You can also summarize the data in R in a customized and repeatable way as well by using |> to pipe the data to the summarize() function. Then tell R what functions to use on which columns. Use min() for the minimum, max() for the maximum, mean() for the mean, median() for the median, sd() for standard deviation, var() for the variance, and n() for the count. Also tell R what to call the output columns and assign new names. We chose these names because they are short as well as relatively easy to both type and understand where the new numbers come from:

```
## R
gb_det_2020_pass_r |>
summarize(min_yac = min(air_yards),
          max_yac = max(air_yards),
          mean_yac = mean(air_yards),
          median_yac = median(air_yards),
          sd_yac = sd(air_yards),
```

```
        var_yac = var(air_yards),
        n_yac = n())
```

Resulting in:

```
# A tibble: 1 × 7
  min_yac max_yac mean_yac median_yac sd_yac var_yac n_yac
    <dbl>   <dbl>    <dbl>      <dbl>  <dbl>   <dbl> <int>
1      NA      NA       NA         NA     NA      NA   181
```

R gives us only NA values. What is going on? Recall that these columns have missing data, so tell R to ignore them by using the na.rm = TRUE option in the functions:

```
## R
gb_det_2020_pass_r |>
summarize(min_yac = min(air_yards, na.rm = TRUE),
          max_yac = max(air_yards, na.rm = TRUE),
          mean_yac = mean(air_yards, na.rm = TRUE),
          median_yac = median(air_yards, na.rm = TRUE),
          sd_yac = sd(air_yards, na.rm = TRUE),
          var_yac = var(air_yards, na.rm = TRUE),
          n_yac = n())
```

Resulting in:

```
# A tibble: 1 × 7
  min_yac max_yac mean_yac median_yac sd_yac var_yac n_yac
    <dbl>   <dbl>    <dbl>      <dbl>  <dbl>   <dbl> <int>
1      -6      50     8.61          5   10.9    120.   181
```

Both Python and R allow you to *group by* variables or calculate statistics by grouping variables such as the mean air_yards for each posteam. Python has a grouping function, groupby(), that can take posteam to calculate the statistics by the possession team (notice Python does not use piping). Instead, string together one function after another. This approach is based on the object-oriented nature of Python compared to the procedural nature of R, both of which have benefits and drawbacks you have to consider:

```
## Python
print(gb_det_2020_pass_py.groupby("posteam").agg(
  {
    "air_yards": ["min", "max", "mean",
                  "median", "std", "var", "count"]
  }
))
```

Resulting in:

```
         air_yards
              min   max      mean median        std        var count
posteam
DET          -6.0  50.0  8.031250    5.0  11.607796 134.740927    32
GB           -4.0  34.0  9.233333    5.0  10.338023 106.874713    30
```

With Python, you can include a second variable by including a second entry in the dictionary. Also, pandas, unlike the tidyverse, allows you to calculate different summaries for each variable by changing the dictionary values:

```Python
## Python
print(gb_det_2020_pass_py.groupby("posteam").agg(
  {
    "yards_after_catch": ["min", "max", "mean",
                          "median", "std", "var", "count"],
    "air_yards": ["min", "max", "mean",
                  "median", "std", "var", "count"]
                }
))
```

Resulting in:

```
          yards_after_catch              ...  air_yards
                        min    max    mean  ...        std        var count
posteam                                    ...
DET                     0.0   20.0  6.900000  ...  11.607796  134.740927    32
GB                     -2.0   19.0  5.555556  ...  10.338023  106.874713    30

[2 rows x 14 columns]
```

R also includes a *group by* function, group_by(), that may be used with piping:

```R
## R
gb_det_2020_pass_r |>
  group_by(posteam) |>
  summarize(min_yac = min(air_yards, na.rm = TRUE),
            max_yac = max(air_yards, na.rm = TRUE),
            mean_yac = mean(air_yards, na.rm = TRUE),
            median_yac = median(air_yards, na.rm = TRUE),
            sd_yac = sd(air_yards, na.rm = TRUE),
            var_yac = var(air_yards, na.rm = TRUE),
            n_yac = n())
```

Resulting in:

```
# A tibble: 3 × 8
  posteam min_yac max_yac mean_yac median_yac sd_yac var_yac n_yac
  <chr>     <dbl>   <dbl>    <dbl>      <dbl>  <dbl>   <dbl> <int>
1 DET          -6      50     8.03          5   11.6    135.    78
2 GB           -4      34     9.23          5   10.3    107.    89
3 <NA>        Inf    -Inf      NaN         NA     NA      NA    14
```

A Note About Presenting Summary Statistics

The key for presenting summary statistics is to make sure you use the information available to you to effectively tell your story. First, knowing your target audience is extremely important. For example, if you're talking to Cris Collinsworth about his next *Sunday Night Football* broadcast (something Eric did at his previous job) or to your buddies at the bar during a game, you're going to present the information differently.

Furthermore, if you're presenting your work to the director of research and strategy for an NFL team, you're probably going to have to supply different—specifically, more—information than in the aforementioned two examples. Likewise, when talking to the director of research and strategy, you will likely need to justify both your estimates and your methodological choices. Conversely, unless you're having beers with Eric and Richard (or other quants), you probably will not be discussing modeling choices over beers!

The *why* is key, and you'll have to dig into data and truly understand it well, so that you can speak it in multiple languages. For example, is the dynamic you're seeing due to coverage differences, the wide receivers, or changes in the quarterback's fundamental ability?

Second, use statistics and modeling to support your story, but do not use them as your entire story. Say, "Drew Brees is still the most accurate passer in football, even after you adjust for situation" rather than "Drew Brees has the highest completion percentage above expected. Period." Adding context to one's work is something that we, as authors, have noticed helps the best quantitative people stand out compared to many quantitative people. In fact, communication skills about numbers helped both of us get our current jobs.

Third, while a picture may be worth a thousand words, walk your reader through the picture. A graph with no context is likely worse than no graph at all because all the graph without context will do is confuse your audience. For a nontechnical audience, you may include a figure and mention the "averages" in your words. Thus, the raw summary statistics may not even be shown in your writing. For more technical audiences, include the details and uncertainty either in text for one or two numbers or in a table or supplemental materials for more summary statistics.

Improving Your Presentation

We have found there are two good ways to improve our presentation of summary statistics. First, present early and present often to people who will give you constructive feedback. Make sure they can understand your message, and if they cannot, ask them what is unclear and figure out how to more clearly make your point. For example, we like to give lectures and seminars to students because we will ask them how they might explain a figure and then they help us to more clearly think about data. Also, if you cannot explain concepts to high school and college students, you do not clearly understand the ideas well.

Second, look at other people's work. Read blogs, read other books, read articles, and watch or listen to podcasts. Other people's examples will help you see what is clear and what is not. Besides casual reading, read critically. What works? What does not work? Why did the authors make a choice? How would you help the author better explain their findings to you? If you have a chance, ask the authors if you see them or interact with them on social media. A diplomatic tweet will likely start a conversation. For example, you might reply to a tweet, *I liked your model output and the insight it gave me to Friday's game. Why did you use X rather than Y?* Conversely, replying to a tweet with *Your model sucked, you should use my favorite model* will likely be ignored or possibly start a pointless flame war and diminish not only the original poster's view of you but also that of other people who read the tweet.

Exercises

1. Repeat the processes within this chapter with a different game.
2. Repeat the processes within this chapter with a different feature, like rushing yards.

Suggested Readings

Many books describe introductory statistics. If you want to learn more about statistics, we suggest reading the first one or two chapters of several books until you find one that *speaks* to you. The following are some books you may wish to consider:

- *Advancing into Analytics: From Excel to Python and R* by George Mount (O'Reilly, 2021). This book assumes you know Excel well but then helps you transition to either R or Python. The book covers the basis of statistics.

- *Statistical Inference via Data Science: A ModernDive into R and the Tidyverse* by Chester Ismay and Albert Y. Kim (CRC Press, 2019), also updated at the book's home page. This book provides a robust introduction to statistical inferences for people who also want to learn R.

- *Practical Statistics for Data Scientists*, 2nd edition, by Peter Bruce et al. (O'Reilly, 2020). This book provides an introduction to statistics for people who already know some R or Python.

- *Introductory Statistics with R* by Peter Dalgaard (Springer 2008) is a classic book; while somewhat dated for code, this was the book Richard used to first learn statistics (and R).

- *Essential Math for Data Science* by Thomas Nield (O'Reilly, 2022) provides a gentle introduction to statistics as well as mathematics for applied data scientists.

Data-Wrangling Fundamentals

Tidy datasets are all alike, but every messy dataset is messy in its own way.

—Hadley Wickham

This appendix focuses on some basics of *data wrangling*, or the process of formatting and cleaning data prior to using it. We include some common but sometimes confusing tools we use on a regular basis. We need a large toolbox, because, as noted by Hadley Wickham, each messy data has its own pathologies. For a more in-depth side-by comparison of Python and R, check out the appendix in *Python and R for the Modern Data Scientist* by Rick J. Scavetta and Boyan Angelov (O'Reilly, 2021).

Data wrangling has many synonyms because almost everybody working with data needs to clean it. Other terms include *data cleaning*, *data formatting*, *data tidying*, *data transformation*, *data manipulation*, *data munging*, and *data mutating*. Basically, people use various terms, so don't be surprised if you see different terms in different sources. Also, in our experience, people inconsistently use these terms. The key take-home is that you'll need to clean, format, transform, or otherwise change your own data at some point. Hence, we included this appendix.

Logic Operators

Logic operators are the same across most languages, including Python and R. The upcoming Table C-1 lists some common operators. Explore these operators by creating a vector in R:

```
## R
score <- c(21, 7, 0, 14)
team <- c("GB", "DEN", "KC", "NYJ")
```

Or, create arrays with numpy in Python:

```Python
## Python
import numpy as np
score = np.array([21, 7, 0, 14])
team = np.array(['GB', 'DEN', 'KC', 'NYJ'])
```

 Python's numpy's arrays differ from base Python's lists and have different behaviors with mathematical functions.

Basic operators are easy to figure out, like > for greater than or < for less than. For example, you can see which elements are greater than 7 in Python:

```Python
## Python
score > 7
```

Resulting in:

```
array([ True, False, False,  True])
```

As you can see, when you use these operators with an array, the operation is performed against each element individually, and all the results are placed into a new array. This is similar to the way the operations work in R, as you will see. For example, with R you can see which elements are less than 15 in R:

```R
## R
score < 15
```

Resulting in:

```
[1] FALSE  TRUE  TRUE  TRUE
```

Less than or equal to, and greater than or equal to, use the equals sign plus the operator: >= is greater than or equal to, and <= is less than or equal to. For example, compare the next code example to the previous one:

```Python
## Python
score <= 14
```

Resulting in:

```
array([False,  True,  True,  True])
```

Other operators are less obvious. Because we already use = to define objects, == is used for *equals*. For example, you can find elements of team that are equal to GB. Make sure you put team in quotes (**"GB"**). Otherwise, the computer thinks you are trying to use an object named GB:

```Python
## Python
team == "GB"
```

Resulting in:

```
array([ True, False, False, False])
```

Using an in-type operator is really useful when you have multiple ways to chart a player playing a similar position. For example, DE (defensive end), OLB (outside linebacker), and ED (edge defender) mean similar things in football, and filtering a dataset for all three terms is often something you do in analysis.

In numpy, you can do this with the .isin() function:

```Python
## Python
position = np.array(['QB', 'DE', 'OLB', 'ED'])
np.isin(position, ['DE', 'OLB', 'ED'])
```

Resulting in:

```
array([False,  True,  True,  True])
```

The pandas package has a similar function for dataframes, covered in "Filtering and Sorting Data" on page 291.

R has a slightly different operator, an %in% function:

```R
## R
position <- c("QB", "DE", "OLB", "ED")
position %in% c("DE", "OLB", "ED")
```

Resulting in:

```
[1] FALSE  TRUE  TRUE  TRUE
```

When using %in%, be careful with the order. For example, compare position %in% c("DE", "OLB", "ED") from the previous example with c("DE", "OLB", "ED") %in% position:

```R
## R
c("DE", "OLB", "ED") %in% position
```

Resulting in:

```
[1] TRUE TRUE TRUE
```

Using in operators can be hard. We will often grab a test subset of our data to make sure our code works as expected. More broadly, do not trust your code until you have convinced yourself that your code works as expected. Casually, use print() statements to peek at your code and make sure it does what you think it is doing. We do this for one-off projects. Formally, unit-testing exists as a method to test code. Python comes with the unittest package, and R has the testthat package for formal testing. We use unit-testing on code we plan to reuse or import code when failure has a large cost.

You can also string together operators by using the *and* operator (&) or the *or* operator (|). Using multiple operators requires the values to be in order as pairs. Our example implies that score corresponds to team. Both vectors are of length 4 in our examples.

For example, you can see which entries are greater than or equal to 7 for score and have a team value of DEN. When working with the numpy arrays, you need to use the where() function, but this logic will be the same and use similar notation with pandas later in this chapter. The results reveal which entry meets the criteria:

```
## Python
np.where((score >= 7) & (team == "DEN"))
```

Resulting in:

```
(array([1]),)
```

You can also use an or operator for a similar comparison to see which values of score are greater than 7 *or* which values of team are equal to DEN:

```
## R
score > 7 | team == "DEN"
```

Resulting in:

```
[1]  TRUE  TRUE FALSE  TRUE
```

You can string together multiple conditions with parentheses. For example, you can see what has score values greater than or equal to 7 *and* team equal to DEN *or* score equal to 0:

```
## Python
np.where((score >= 7) & (team == "DEN") | (score == 0))
```

Resulting in:

```
(array([1, 2]),)
```

Likewise, similar notation may be used in R:

```
## R
(score >= 7 & team == "DEN") | (score == 0)
```

Resulting in:

```
[1] FALSE   TRUE   TRUE FALSE
```

Table C-1. Common logical operators. [a]

Symbol	Example	Name	Question
==	score == 2	Equals	Is score equal to 2?
!=	score != 2	Not equals	Is score not equal to 2?
>	score > 2	Greater than	Is score greater than 2?
<	score < 2	Less than	Is score less than 2?
>=	score >= 2	Greater than or equal to	Is score greater than or equal to 2?
<=	score <= 2	Less than or equal to	Is score less than or equal to 2?
|	(score > 2) | (team =="GB")	Or	Is score less than 2, or team equal to GB?
&	(score > 2) & (team =="GB")	And	Is score less than 2, and team equal to GB?

[a] pandas sometimes uses ~ rather than ! for *not* in some situations.

Filtering and Sorting Data

In the previous section, you learned about logical operators. These functions serve as the foundation of filtering data. In fact, when we get stuck with filtering, we often build small test cases like the ones in "Logic Operators" on page 287 to make sure we understand our data and the way our filters work (or, as is sometimes the case, do not work).

Filtering can be hard. Start small and build complexity into your filtering commands. Keep adding details until you are able to solve your problem. Sometimes you might need to use two or more smaller filters rather than one grand filter operation. This is OK. Get your code working before worrying about optimization.

You will work with the Green Bay–Detroit data from the second week of the 2020 season. First, read in the data and do a simple filter to look at plays that had a yards-after-catch value greater than 15 yards to get an idea of where some big plays were generated.

In R, load the `tidyverse` and `nflfastR` packages and then load the data for 2020:

```
## R
library(tidyverse)
library(nflfastR)

# Load all data
pbp_r <- load_pbp(2020)
```

In Python, import the `pandas`, `numpy`, and `nfl_data_py` packages and then load the data for 2020:

```
## Python
import pandas as pd
import numpy as np
import nfl_data_py as nfl

# Load all data

pbp_py = nfl.import_pbp_data([2020])
```

Resulting in:

```
2020 done.
Downcasting floats.
```

In R, use the `filter()` function next. The first argument into filter is `data`. The second argument is the `filter` criteria. Filter out the Detroit at Green Bay game and select some passing columns:

```
# Filter out game data
gb_det_2020_r_pass <-
    pbp_r |>
    filter(home_team == 'GB' & away_team == 'DET') |>
    select(posteam, yards_after_catch, air_yards,
           pass_location, qb_scramble)
```

Next, `filter()` the plays with `yards_after_catch` that were greater than 15:

```
gb_det_2020_r_pass |>
filter(yards_after_catch > 15)
```

Resulting in:

```
# A tibble: 5 × 5
  posteam yards_after_catch air_yards pass_location qb_scramble
  <chr>               <dbl>     <dbl> <chr>               <dbl>
1 DET                    16        13 left                    0
2 GB                     19         3 right                   0
3 GB                     19         6 right                   0
4 DET                    16         1 middle                  0
5 DET                    20        16 middle                  0
```

With R and Python, you do not always need to use argument names. Instead, the languages match arguments with their predefined order. This order is listed in the help files. For example, with `gb_det_2020_r_pass |> filter(yards_after_catch > 15)`, you could have written `gb_det_2020_r_pass |> filter(filter = yards_after_catch > 15)`. We usually define argument names for more complex functions or when we want to be clear. It is better to err on the side of being explicit and use the argument names, because doing this makes your code easier to read.

Notice in this example that plays that generated a lot of yards after the catch come in many shapes and sizes, including short throws with 1 yard in the air, and longer throws with 16 yards in the air. You can also filter with multiple arguments by using the "and" operator, &. For example, you can filter by yards after catch being greater than 15 and Detroit on offense:

```
## R
gb_det_2020_r_pass  |>
filter(yards_after_catch > 15 & posteam == "DET")
```

Resulting in:

```
# A tibble: 3 × 5
  posteam yards_after_catch air_yards pass_location qb_scramble
  <chr>               <dbl>     <dbl> <chr>               <dbl>
1 DET                    16        13 left                    0
2 DET                    16         1 middle                  0
3 DET                    20        16 middle                  0
```

However, what if you want to look at plays with yards after catch being greater than 15 yards or air yards being greater than 20 yards and Detroit the offensive team? If you try `yards_after_catch > 15 | air_yards > 20 & posteam == "DET"` in the filter, you get results with both Green Bay and Detroit rather than only Detroit. This is because the order of the operations is different than you intended:

```
## R
gb_det_2020_r_pass |>
filter(yards_after_catch > 15 | air_yards > 20 &
       posteam == "DET")
```

Resulting in:

```
# A tibble: 9 × 5
  posteam yards_after_catch air_yards pass_location qb_scramble
  <chr>               <dbl>     <dbl> <chr>               <dbl>
1 DET                    16        13 left                    0
2 GB                     19         3 right                   0
3 DET                    NA        28 left                    0
4 DET                    NA        28 right                   0
5 GB                     19         6 right                   0
```

```
 6 DET                    16           1 middle              0
 7 DET                     0          24 right               0
 8 DET                    20          16 middle              0
 9 DET                    NA          50 left                0
```

You get all plays with yards after catching being greater than 15 or all plays with yards greater than 20 and Detroit starting with possession of the ball. Instead, add a set of parentheses to the filter: `(yards_after_catch > 15 | air_yards > 20) & posteam == "DET"`.

 The *order of operations* refers to the way people perform math functions and computers evaluate code. The key takeaway is that both order and grouping of functions changes the output. For example, $1 + 2 \times 3 = 1 + 6 = 7$ is different from $(1 + 2) \times 3 = 3 \times 3 = 9$. When you combine operators, the default order of operations sometimes leads to unexpected outcomes, as in the previous example you expected to filter out GB but did not. To avoid this type of confusion, parentheses help you explicitly choose the order that you intend.

The use of parentheses in both coding and mathematics align, so the order of operations starts with the innermost set of parentheses and then moves outward:

```
## R
gb_det_2020_r_pass |>
filter((yards_after_catch > 15 | air_yards > 20) &
       posteam == "DET")
```

Resulting in:

```
# A tibble: 7 × 5
  posteam yards_after_catch air_yards pass_location qb_scramble
  <chr>              <dbl>     <dbl> <chr>               <dbl>
1 DET                   16        13 left                    0
2 DET                   NA        28 left                    0
3 DET                   NA        28 right                   0
4 DET                   16         1 middle                  0
5 DET                    0        24 right                   0
6 DET                   20        16 middle                  0
7 DET                   NA        50 left                    0
```

You can also change the filter to look at only possession teams that are not Detroit by using the "not-equal-to" operator, `!=`. In this case, the "not-equal-to" operator gives you Green Bay's admissible offensive plays, but this would not always be the case. For example, if you were working with season long data with all teams, the "not-equal-to" operator would give you data for the 31 other NFL teams:

```
## R
gb_det_2020_r_pass |>
    filter((yards_after_catch > 15 | air_yards > 20) &
        posteam != "DET")
```

Resulting in:

```
# A tibble: 8 × 5
   posteam yards_after_catch air_yards pass_location qb_scramble
   <chr>               <dbl>     <dbl> <chr>               <dbl>
1 GB                     NA        26 left                    0
2 GB                     NA        25 left                    0
3 GB                     19         3 right                   0
4 GB                     NA        24 right                   0
5 GB                      4        26 right                   0
6 GB                     NA        28 left                    0
7 GB                     19         6 right                   0
8 GB                      7        34 right                   0
```

In Python with pandas, filtering is done with similar logical structure to the tidy
verse in R, but with different syntax. First, Python uses a .query() function. Second,
the logical operator is inside quotes:

```
## Python
gb_det_2020_py_pass = \
    pbp_py\
        .query("home_team == 'GB' & away_team == 'DET'")\
        [["posteam", "yards_after_catch","air_yards",
            "pass_location", "qb_scramble"]]

print(gb_det_2020_py_pass.query("yards_after_catch > 15"))
```

Resulting in:

```
      posteam yards_after_catch air_yards pass_location qb_scramble
4034      DET              16.0      13.0          left         0.0
4077       GB              19.0       3.0         right         0.0
4156       GB              19.0       6.0         right         0.0
4171      DET              16.0       1.0        middle         0.0
4199      DET              20.0      16.0        middle         0.0
```

Notice that the or operator, |, works the same with both languages:

```
## Python
print(gb_det_2020_py_pass.query("yards_after_catch > 15 | air_yards > 20"))
```

Resulting in:

```
      posteam yards_after_catch air_yards pass_location qb_scramble
4034      DET              16.0      13.0          left         0.0
4051       GB               NaN      26.0          left         0.0
4055       GB               NaN      25.0          left         0.0
4077       GB              19.0       3.0         right         0.0
4089      DET               NaN      28.0          left         0.0
4090      DET               NaN      28.0         right         0.0
```

4104	GB	NaN	24.0	right	0.0
4138	GB	4.0	26.0	right	0.0
4142	GB	NaN	28.0	left	0.0
4156	GB	19.0	6.0	right	0.0
4171	DET	16.0	1.0	middle	0.0
4176	DET	0.0	24.0	right	0.0
4182	GB	7.0	34.0	right	0.0
4199	DET	20.0	16.0	middle	0.0
4203	DET	NaN	50.0	left	0.0

 In R or Python, you can use single quotes (') or double quotes (").
When using functions such as .query() in Python, you see why
the languages contain two approaches for quoting. You could use
"posteam == 'DET'" or 'posteam == "DET"'. The languages do
not care if you use single or double quotes, but you need to be
consistent within the same function call.

In Python, when your code gets too long to easily read on a line, you need a backslash
(\), for Python to understand the line break. This is because Python treats whitespace
as a special type of code, whereas R usually treats *whitespace*, such as spaces, indenta-
tions, or line breaks, simply as aesthetic. To a novice, this part of Python can be
frustrating, but the use of whitespace is a beautiful part of the language once you gain
experience to appreciate it.

Next, look at the use of parentheses with the or operator and the and operator, just as
in R:

```
print(gb_det_2020_py_pass.query("(yards_after_catch > 15 | \
                                  air_yards > 20) & \
                                  posteam == 'DET'"))
```

Resulting in:

	posteam	yards_after_catch	air_yards	pass_location	qb_scramble
4034	DET	16.0	13.0	left	0.0
4089	DET	NaN	28.0	left	0.0
4090	DET	NaN	28.0	right	0.0
4171	DET	16.0	1.0	middle	0.0
4176	DET	0.0	24.0	right	0.0
4199	DET	20.0	16.0	middle	0.0
4203	DET	NaN	50.0	left	0.0

Cleaning

Having accurate data is important for sports analytics, as the edge in sports like
football can be as little as one or two percentage points over your opponents, the
sportsbook, or other players in fantasy football. Cleaning data by hand via programs
such as Excel can be tedious and leaves no log indicating which values were changed.

Also, fixing one or two systematic errors by hand can easily be done with Excel, but fixing or reformatting thousands of cells in Excel would be difficult and time-consuming. Luckily, you can use scripting to help you clean data.

 When estimating which team will win a game, the *edge* refers to the difference between the predictor's estimated probability and the market's estimated probability (plus the book's commission, or vigorish). For example, if sportsbooks are offering the Minnesota Vikings at a price of 2–1 to win a game against the Green Bay Packers in Lambeau Field, they are saying that to bet the Vikings, you need to believe that they have more than a 1 / (2 + 1) x 100% = 33.3% chance to win the game. If you make Vikings 36% to win the game, you have a 3% edge betting the Vikings. As information and the synthesizing of information have become more prevalent, edges have become smaller (as the markets have become more efficient). Professional bettors are always in search of better data and better ways to synthesize data, to outrun the increasingly efficient markets they play in.

Consider this example dataframe in `pandas`:

```
wrong_number = \
    pd.DataFrame({"col1": ["a", "b"],
                  "col2": ["10", "12"],
                  "col3": [2, 44]})
```

Notice that `col2` has a 1O ("one-oh") rather than a 10 ("one-zero," or ten). These types of mistakes are fairly common in hand-entered data. This may be fixed using code.

 Both R and Python allow you to access dataframes by using a coordinate-like system with rows as the first entry and columns as the second entry. Think of this like a game of Battleship or Bingo, when people call out cells like *A4* or *B2*. The `pandas` package has `.loc[]` to access rows or columns by names. For example, to access the first value in the `posteam` column of the play-by-play data, run `pbp_py.loc[1, "posteam"]` (1 is the row name or index, and `posteam` is the column name). To access the first row of the first column, run `print(pbp_py.iloc[1, 0])`. Compare these two methods. What column is 0? It is better to use filters or explicit names. This way, if your data changes, you call the correct cell. Also, this way, future you and other people will know why you are trying to access specific cells.

Use the locate function, `.loc()`, to locate the wrong cell. Also, select the column, `col2`. Last, replace the wrong value with a 10 (ten):

```
## Python
wrong_number.loc[wrong_number.col2 == "10", "col2"] = 10
```

Look at the dataframe's information, though, and you will see that col2 is still an object rather than a number or integer:

```
## Python
wrong_number.info()
```

Resulting in:

```
<class 'pandas.core.frame.DataFrame'>
RangeIndex: 2 entries, 0 to 1
Data columns (total 3 columns):
 #   Column  Non-Null Count  Dtype
---  ------  --------------  -----
 0   col1    2 non-null      object
 1   col2    2 non-null      object
 2   col3    2 non-null      int64
dtypes: int64(1), object(2)
memory usage: 176.0+ bytes
```

 Both R and Python usually require users to save data files as outputs after editing (a counter-example being the inplace=True option in some pandas functions). Otherwise, the computer will not save your changes. Failure to update or save objects can cost you hours of debugging code, as we have learned from our own experiences.

Change this by using the to_numeric() function from pandas and then look at the information for the dataframe. Next, save the results to col2 and rewrite the old data. If you skip this step, the computer will not save your edits:

```
## Python
wrong_number["col2"] = \
    pd.to_numeric(wrong_number["col2"])
wrong_number.info()
```

Resulting in:

```
<class 'pandas.core.frame.DataFrame'>
RangeIndex: 2 entries, 0 to 1
Data columns (total 3 columns):
 #   Column  Non-Null Count  Dtype
---  ------  --------------  -----
 0   col1    2 non-null      object
 1   col2    2 non-null      int64
 2   col3    2 non-null      int64
dtypes: int64(2), object(1)
memory usage: 176.0+ bytes
```

Notice that now the column has been changed to an integer.

If you want to save these changes for later, you can use the to_csv() function to save the outputs. Generally, you will want to use a new filename that makes sense to you now, to others, and to your future self. Because the dataframe does not have meaningful row names or an index, tell pandas to not save this information by using index=False:

```Python
## Python
wrong_number.to_csv("wrong_number_corrected.csv", index = False)
```

R uses slightly different syntax. First, use the mutate() function to change the column. Next, tell R to change col2 by using col2 = ... Then use the ifelse() function to tell R to change col2 if it is equal to 10 ("one-oh") to be 10 ("one-zero," or ten), or to use the current value in col2:

```R
## R
wrong_number <-
  tibble(col1 = c("a", "b"),
         col2 = c("10", "12"),
         col3 = c(2, 44))
wrong_number <-
  wrong_number |>
  mutate(col2 = ifelse(col2 == "10", 10, col2))
```

Next, just as in Python, change col2 to be numeric. In R, use the as.numeric() function. Then look at the dataframe structure by using str():

```R
## R
wrong_number <- mutate(wrong_number, col2 = as.numeric(col2))
str(wrong_number)
```

Resulting in:

```
tibble [2 × 3] (S3: tbl_df/tbl/data.frame)
 $ col1: chr [1:2] "a" "b"
 $ col2: num [1:2] 10 12
 $ col3: num [1:2] 2 44
```

Finally, just as in Python, save the file by using a name that makes sense to both the current you and future you. Hopefully, this name makes sense to other people. Creating names can be one of the most difficult parts of programming. With R, use the write_csv() function:

```R
## R
write_csv(x = wrong_number,
          file = "wrong_numbers_corrected.csv")
```

 Python uses False for the logical false and True for the logical true. R uses FALSE for false and TRUE for true. If you are switching between the languages, be careful with these terms.

Piping in R

With programming, sometimes you want to pass outputs from one function to another without needing to save the intermediate outputs. In mathematics, this is called *composition*, and while teaching college math classes, Eric observed this to be one of the more misunderstood procedures because of the confusing notation. In computer programming, this is called *piping* because outputs are piped from one function to another.

Luckily, R has allowed composition through the piping operators with the `tidyverse` that has a pipe function, `%>%`. As of R version 4.1, released in 2021, base R also now includes a `|>` pipe operator. We use the base R pipe operator in this book, but you may see both "in the wild" when looking at other people's code or websites.

 The `tidyverse` pipe allows piping to any function's input option by using a period. This period is optional with the `tidy verse` pipe. And the `tidyverse` pipe will, by default, use the first function input with piping. For example, you might code `read_csv("my_file.scv") %>% func(x = col1, data = .)`, or `read_csv("my_file.scv") %>% function(col1)`. With `|>`, you can pass to only the first input; thus you would need to define all inputs prior to the one you are piping (in this case, `data`). With the `|>` pipe, your code would be written as `read_csv("my_file.scv") |> function(x = col1)`.

 Any reference material can become dated, especially online tutorials. The piping example demonstrates how any tutorial created before R 4.1 would not include the new piping notation. Thus, when using a tutorial, examine when the material was written and ensure that you can re-create a tutorial before applying it to your problem. And, when using sites such as StackOverflow (*https://stackoverflow.com*), we look at several of the top answers and questions to make sure the accepted answer has not become outdated as languages change. The best answer in 2013 may not be the best answer in 2023.

We cover piping here for two reasons. First, you will likely see it when you start to look at other people's code as you teach yourself. Second, piping allows you to be more efficient with coding once you get the hang of it.

Checking and Cleaning Data for Outliers

Data often contains errors. Perhaps people collecting or entering the data made a mistake. Or, maybe an instrument like a weather station malfunctioned. Sometimes, computer systems corrupt or otherwise change files. In football, quite often there will be errors in things like number of air yards generated, yards after the catch earned, or even the player targeted. Resolving these errors quickly, and often through data wrangling, is a required process of learning more about the game. Chapter 2 presented tools to help you catch these errors.

You'll go through and find and remove an outlier with both languages. Revisiting the wrong_number dataframe, perhaps col3 should be only single digits. The summary() function would help you see this value is wrong in R:

```
## R
wrong_number |>
  summary()
```

Resulting in:

```
      col1              col2            col3
 Length:2          Min.   :10.0    Min.   : 2.0
 Class :character  1st Qu.:10.5    1st Qu.:12.5
 Mode  :character  Median :11.0    Median :23.0
                   Mean   :11.0    Mean   :23.0
                   3rd Qu.:11.5    3rd Qu.:33.5
                   Max.   :12.0    Max.   :44.0
```

Likewise, the describe() function in Python would help you catch an outlier:

```
## Python
wrong_number.describe()
```

Resulting in:

```
            col2        col3
count   2.000000    2.000000
mean   11.000000   23.000000
std     1.414214   29.698485
min    10.000000    2.000000
25%    10.500000   12.500000
50%    11.000000   23.000000
75%    11.500000   33.500000
max    12.000000   44.000000
```

Using the tools covered in the previous section, you can remove the outlier.

Merging Multiple Datasets

Sometimes you will need to combine datasets. For example, often you will want to adjust the results of a play—say, the number of passing yards—by the weather in which the game was played. Both `pandas` and the `tidyverse` readily allow merging datasets. For example, perhaps you have team and game data you want to merge from both datasets. Or, maybe you want to merge weather data to the play-by-play data.

For this example, create two dataframes and then merge them. One dataframe will be city information that contains the teams' names and cities. The other will be a schedule. We have you create a small example for multiple reasons. First, a small toy dataset is easier to handle and see, compared to a large dataset. Second, we often create toy datasets to make sure our merges work.

 When learning something new (like merging dataframes), start with a small example you understand. The small example will be easier to debug, fail faster, and understand compared to a large example or actual dataset.

You might be wondering, why merge these dataframes? We often have to do merges like this when summarizing data because we want or need a prettier name. Likewise, we often need to change names for plots. Next, you may be wondering, why not type these values into a spreadsheet? Manually typing can be tedious and error prone. Plus, doing tens, hundreds, or even thousands of games would take a long time to type.

As you create the dataframes in R, remember that each column you create is a vector:

```
## R
library(tidyverse)
city_data <- data.frame(city = c("DET", "GB", "HOU"),
                        team = c("Lions", "Packers", "Texans"))
schedule <- data.frame(home = c("GB", "DET"),
                       away = c("DET", "HOU"))
```

As you create the dataframes in Python, remember that the `DataFrame()` function uses a dictionary to create columns and elements in the columns:

```
## Python
import pandas as pd
city_data = \
    pd.DataFrame({"city" : ["DET", "GB", "HOU"],
                  "team" : ["Lions", "Packers", "Texans"]})
schedule = \
    pd.DataFrame({"home" : ["GB", "DET"],
                  "away" : ["DET", "HOU"]})
```

Now that you have the datasets, use them to explore various merges. Both pandas and the tidyverse base their merge functions on SQL. The join functions require a common, shared key or multiple keys between the two dataframes. In the tidyverse, this argument is called *by*—for example, joining city and schedule dataframes by *team name* and *home* team columns. In pandas, this argument is called *on*—for example, joining city and schedule dataframes on *team name* and *home* team columns.

We use four main joins on a regular basis, and these are included with the tidyverse and pandas. The pandas package has both a merge() and a join() function. The merge() function contains almost everything that join() does, plus some more, so we will include only merge() here. With both Python and R, there are two datasets, a left one and a right one. The *left dataset* is the one on the left (or the first dataset), and the *right dataset* is the one on the right (or the second dataset).

For the example, you want to create a new dataframe that includes both the schedule and the teams' names. Use this to explore the various types of joins. Think of this example as the fairy tale of Goldilocks and the four joins (based on the original story of *Goldilocks and the Three Bears* (*https://oreil.ly/iHnci*)). Rather than a girl trying the bears' beds and food, you'll be exploring data joins, listed in Table C-2. This problem has two steps. The first step is to add in the home team's name. The second step is to add in the away team's name. At the end, we will show you the complete workflow because it also involves renaming columns.

 Football analytics, like the broader field of data science, usually involves breaking big jobs into smaller jobs. As you become more experienced, you will become better at seeing the small steps and knowing where and how to reuse them. When faced with intimidating problems, we break them into smaller steps that we can readily solve. Often our first step is to write or draw out our coding needs, much as you may have outlined a paper in high school or college before writing the paper.

First, examine a *full*, or *outer, join*. This merges both dataframes based on all values in both dataframes' keys. If one or both keys contain values not found in the other dataset, these are replaced by missing values (NA in R, NaN in Python). For both languages, schedule will be your left dataframe, and city_data will be your right dataframe. Because both dataframes do not have the same key (or, specifically, the column with the same names), the computer needs to know how to pair up the keys (specifically, which columns link the two dataframes).

In R, use the full_join() function. Put schedule in first, followed by city_data. Tell R to *join* the dataframes by using home as the left key matching up with city as the right key:

```
## R
print(
  full_join(schedule, city_data,
  by = c("home" = "city"))
  )
```

Resulting in:

```
  home away    team
1   GB  DET Packers
2  DET  HOU   Lions
3  HOU <NA>  Texans
```

Notice that you get three entries because the `city_data` has three rows. The missing value is replaced by NA. Notice that R dropped the duplicate column and has only three columns.

In Python, use the `.merge()` function on the `schedule` dataframe. And notice that `schedule` is on the left. The first argument is `city_data`. Tell pandas how to merge—specifically, an `outer` merge. Then tell pandas to use `home` as the left key and `city` as the right key:

```
## Python
print(schedule.merge(city_data, how = "outer",
                     left_on = "home", right_on = "city"))
```

Resulting in:

```
  home away city    team
0   GB  DET   GB Packers
1  DET  HOU  DET   Lions
2  NaN  NaN  HOU  Texans
```

Notice `pandas` kept all four columns. Also note that both `home` and `away` are NaN for the new dataframe.

 This example demonstrates how Python tends to be an object-orientated language, and R tends to be a functional language. Python uses `.merge()` as an object contained by the dataframe `schedule`. R uses a `full_join()` as a function on two objects, `schedule` and `city_data`. Although R and Python both contain object-orientated and functional features, this example nicely demonstrates the underlying philosophies of the two languages.

Think of this distinction of language types similar to the way some football teams are built for a run offense and others for a pass offense. Under certain circumstances, one language can be better than the other, but usually both contain the tools for a given job. Advanced data scientists recognize these trade-offs between languages and will switch languages to fit their needs.

Next, do an *inner join*. This joins only the shared key values. Whereas an outer join may possibly grow dataframes, an inner join shrinks dataframes. The R syntax is very similar to the previous example; only the function name changes. However, notice that the output has only three values:

```
## R
print(inner_join(schedule, city_data, by = c("home" = "city")))
```

Resulting in:

```
  home away     team
1   GB  DET  Packers
2  DET  HOU    Lions
```

Like R, the Python code is similar. In Python, use the same function, but a different how argument:

```
## Python
print(schedule.merge(city_data, how = "inner",
                     left_on = "home", right_on = "city"))
```

Resulting in:

```
  home away city     team
0   GB  DET   GB  Packers
1  DET  HOU  DET    Lions
```

Next, do a *right join*. The right join keeps all the values from the right dataframe. For this specific case, the outputs are the same as the outer join. This is an artifact of the example and may not always be the case. With R, simply change the function name to right_join():

```
## R
print(right_join(schedule, city_data, by = c("home" = "city")))
```

Resulting in:

```
  home away     team
1   GB  DET  Packers
2  DET  HOU    Lions
3  HOU <NA>   Texans
```

With Python, change how to right:

```
## Python
print(schedule.merge(city_data, how = "right",
                     left_on = "home", right_on = "city"))
```

Resulting in:

```
  home away city     team
0  DET  HOU  DET    Lions
1   GB  DET   GB  Packers
2  NaN  NaN  HOU   Texans
```

A *left join* is the opposite of a right join. This keeps all the values from the left dataframe. In fact, rather than switching the function, you could switch the order of inputs. Consider merging dataframes A and B in Python that share a common column, key:

```Python
## Python
A.merge(B, how = "left", on = "key")
```

This could also be written in reverse:

```Python
## Python
B.merge(A, how = "right", on = "key")
```

Here is what the R code and output look like:

```R
## R
print(left_join(schedule, city_data, by = c("home" = "city")))
```

Resulting in:

```
  home away    team
1   GB  DET Packers
2  DET  HOU   Lions
```

The Python code also looks similar to the right join. For both outputs, the left join was the same as the inner join. This is an artifact of the example choice and will not always be the case. Here, the left dataframe had fewer rows than the right dataframe. Hence, this occurred in the example:

```Python
## Python
print(schedule.merge(city_data, how = "left",
                     left_on = "home", right_on = "city"))
```

Resulting in:

```
  home away city    team
0   GB  DET   GB Packers
1  DET  HOU  DET   Lions
```

Table C-2. Common join types in R and Python.

Name	Brief description	tidyverse function	pandas `merge()` syntax
Full/outer join	Merges based on all key values	`full_join(left_data, right_data)`	`left_data.merge(right_data, how = "outer"`
Inner join	Merges based only on shared key values	`inner_join(left_data, right_data)`	`left_data.merge(right_data, how = "inner"`
Left join	Merges based only on *left* data's key values	`left_join(left_data, right_data)`	`left_data.merge(right_data, how = "left"`
Right join	Merges based only on *right* data's key values	`right_join(left_data, right_data)`	`left_data.merge(right_data, how = "right"`

Let's return to the initial problem: "How do you merge the dataframe to include the team names for both the home and away teams?"

Multiple solutions exist, as is often the case with programming. We use multiple left joins because we think about adding data to a schedule and putting this dataframe on the left. However, you might think about the problem differently, which is OK. In fact, you might be able to think about and come up with a better way to do this that is either quicker, easier to read, or uses less code.

 Unlike high school math, both statistics and coding often have no single best or right way to do something. Instead, many unique solutions exist. Some people play a game called *code golf*, in which they try to solve a problem by using the fewest lines of code; see, for example, the Stack Exchange Code Golf page (*https://code golf.stackexchange.com*). But the fewest lines of code is usually not the best answer in real life. Instead, focus on writing code that you and other people can read later. Also new tools such as GitHub's Copilot can help you see and compare methods for coding the same task.

So, we will use a series of left joins (although we could also do everything in reverse, using right joins). Here is our step-by-step solution:

1. Merge in for the home team.
2. Rename column in R, rename and delete column in Python.
3. Merge in for away team. This step is needed for clarity and to avoid duplicate names.
4. Rename columns in R; rename and delete columns in Python.
5. Make sure the output is saved to a new dataframe, `schedule_name`.

The following are some notes about how and why we use these specific steps. Whether we merged by the away or home order is not important, and we arbitrarily selected order. We needed to rename columns to avoid duplicate names later and to keep column names clear. The importance of this will become evident when you have to clean up your own mess or somebody else's messy code! Lastly, we encourage you to start with one line of code and keep adding more code until you understand the big picture. That's how we constructed this example.

With the R example, use piping to avoid rewriting objects as you did for the Python example. First, take the `schedule` dataframe and then left-join to the `city_data`. Tell R to join by (or match) the `home` column to the `city` column. Then rename the `team` column to `home_team`. This helps us keep the team columns straight in the final dataframe. Then repeat these steps and join the away team data:

```
## R
schedule_name <-
    schedule |>
    left_join(city_data, by = c("home" = "city"))  |>
    rename(home_team = team)  |>
    left_join(city_data, by = c("away" = "city"))  |>
    rename(away_team = team)
print(schedule_name)
```

Resulting in:

```
  home away home_team away_team
1   GB  DET   Packers     Lions
2  DET  HOU     Lions    Texans
```

With Python, create temporary objects rather than piping. This is because the pandas piping is not as intuitive to us and requires writing custom functions, something beyond the scope of this book. Furthermore, some people like writing out code to see all the steps, and we want to show you a second technique for this example.

In Python, first do a left merge. Tell Python we use home for the left merge on and city for the right merge on. Then rename the team column to home_team. The pandas rename() function requires a dictionary as an input. Then, tell pandas to remove, or .drop(), the city column to avoid confusion later. Then repeat these steps for the away team:

```
## Python
step_1 = schedule.merge(city_data, how = "left",
                        left_on = "home", right_on = "city")
step_2 = step_1.rename(columns =
                    {"team": "home_team"}).drop(columns = "city")
step_3 = step_2.merge(city_data, how = "left",
                      left_on = "away", right_on = "city")
schedule_name = step_3.rename(columns =
                        {"team": "home_team"}).drop(columns = "city")
print(schedule_name)
```

Resulting in:

```
  home away home_team home_team
0   GB  DET   Packers     Lions
1  DET  HOU     Lions    Texans
```

Glossary

air yards

The distance traveled by the pass from the line of scrimmage to the intended receiver, whether or not the pass was complete.

average depth of target (aDOT)

The average air yards traveled on targeted passes for quarterbacks and targets for receivers.

binomial

A type of statistical distribution with a binary (0/1-type) response. Examples include wins/losses for a game or sacks/no-sacks from a play.

bins

The discrete categories of numbers used to summarize data in a histogram.

book

Short for *sportsbook* in football gambling. This is the person, group, casino, or other similar enterprise that takes wagers on sporting (and other) events.

bounded

Describes a number whose value is constrained by other values. For example, a percentage is bounded by 0% and 100%.

boxplots

A data visualization that uses a "box" to show the middle 50% of data and stems for the upper and lower quartiles. Commonly, outliers are plotted as dots. The plots are also known as *box-and-whisker plots*.

Cheeseheads

Fans of the Green Bay Packers. For example, Richard is a Cheesehead because he likes the Packers. Eric is not a fan of the Packers and is therefore not a Cheesehead.

closing line

The final price offered by a sportsbook before a game starts. In theory, this contains all the opinions, expressed through wagers, of all bettors who have enough influence to move the line into place.

clustering

A type of statistical method for dividing data points into similar groups (*clusters*) based on a set of features.

coefficient

The predictor estimates from a regression-type analysis. Slopes and intercepts are special cases of coefficients.

completion percentage over expected (CPOE)

The rate at which a quarterback has successful (*completed*) passes, compared to what would be predicted (*expected*) given a situation based on an expected completion percentage model.

confidence interval (CI)

A measure of uncertainty around a point estimate such as a mean or regression coefficient. For example, a 95% CI around a mean will contain the true mean 95% of the time if you repeat the observation process many, many, many times. But, you will never know which 5% of times you are wrong.

context

What is going on around a situation; the factors involved in a play, such as the down, yards to go, and field position.

controlled for

Including one or more extra variables in a regression or regression-like model. For example, pass completion might be controlled for yards to go. See also *corrected for* and *normalized*.

corrected for

A synonym for *controlled for*.

data dictionary

Data about data. Also a synonym for *metadata*.

data pipeline

The flow of data from one location to another, with the data undergoing changes such as formatting along the way. See also *pipe*.

data wrangling

The process of getting data into the format you need to solve your problems. Synonyms include *data cleaning*, *data formatting*, *data tidying*, *data transformation*, *data manipulation*, *data munging*, and *data mutating*.

degrees of freedom

The "extra" number of data points left over from fitting a model.

dimensionality reduction

A statistical approach for reducing the number of features by creating new, independent features. Principal component analysis (PCA) is an example of one type of dimensionality reduction.

dimensions (of data)

The number of variables needed to describe data. Graphically, this is the number of axes needed to describe the data. Tabularly, this is the number of columns needed to describe the data. Algebraically, this is the number of independent variables needed to describe the data.

distance

The number of yards remaining to either obtain a new first down or score a touchdown.

down

A finite number of plays to advance the football a certain distance (measured in yards) and either score or obtain a new set of plays before a team loses possession of the ball.

draft approximate value (DrAV)

The approximate value generated by a player picked for his drafting team. This is a metric developed by Pro Football Reference.

draft capital

The resources a team uses during the NFL Draft, including the number of picks, pick rounds, and pick numbers.

edge

An advantage over the betting markets for predicting outcomes, usually expressed as a percentage.

expected point

The estimated, or *expected*, value for the number of *points* one would expect a team to score given the current game situation on that drive.

expected points added (EPA)

The difference between a team's expected points from one play to the next, measuring the success of the play.

exploratory data analysis (EDA)

A subset of statistical analysis that analyzes data by describing or summarizing its main characteristics. Usually, this involves both graphical summaries such as plots and numerical summaries such as means and standard deviations.

feature

A predictor variable in a model. This term is used more commonly by data scientists whereas statisticians tend to use *predictor* or *dependent variable*.

for loop

A computer programming tool that repeats (or *loops*) over a function *for* a predefined number of iterations.

generalized linear models (GLMs)

An extension of linear models (such as simple linear regression and multiple regression) to include a link function and non-normal response variable such as logistic regression with binary data or Poisson regression with count data.

gridiron football

A synonym for American football.

group by

A concept from SQL-type languages that describes taking data and creating subgroups (or *grouping*) based on (or *by*) a variable. For example, you might take Aaron Rodger's passing yards and *group by* season to calculate his average passing yards per season.

handle

The total amount of money placed by bettors across all markets.

high-leverage situations

Important plays that determine the outcome of games. For example, converting the ball on third down, or fourth and goal. These plays, while of great importance, are generally not predictive game to game or season to season.

histogram

A type of plot that summarizes counts of data into discrete bins.

hit

Two uses in this book. A football colliding with another is a *hit*. Additionally, a computer script trying to download from a web page *hits* the page when trying to download.

interaction

In a regression-type model, sometimes two predictors or features change together. A relation (or *interaction*) between these two terms allows this to be included within the model.

intercept

The point where a simple linear regression crosses through 0. Also, sometimes used to refer to multiple regression coefficients with a discrete predictor variable.

internal

For sports bettors, the price they would offer the game if they were a sportsbook. The discrepancy between this value and the actual price of the game determines the edge.

interquartile range

The difference between the first and third quartile. See also *quartile*.

lag

A delay or offset. For example, comparing the number of passes per quarterback per game in one season (such as 2022) to the previous season (such as 2021) would have a lag of 1 season.

link function

The function that maps between the observed scale and model scale in a generalized linear model. For example, a logit function can link between the observed probability of an outcome occurring on the bounded 0–1 probability scale to the log-odds scale that ranges from negative to positive infinity.

log odds

Odds on the log scale.

long pass

A pass typically longer than 20 yards, although the actual threshold may vary.

metadata

The data describing the data. For example, metadata might indicate whether a time column displays minutes, seconds, or decimal minutes. This can often be thought of as a synonym for *data dictionary*.

mixed-effect model

A model with both fixed effects and random effects. See also *random-effect model*. Synonyms include *hierarchical model*, *multilevel model*, and *repeated-measure* or *repeated-observation model*.

moneyline

In American football, a bet on a team winning straight up.

multiple regression

A type of regression with more than one predictor variable. Simple linear regression is a special type of multiple regression.

normalize

This term has multiple definitions. In the book, we use it to refer to accounting for other variables in regression analysis. Also see *correct for* or *control for*. Normalization may also be used to define a transformation of data. Specifically, data are transformed to be on a normal distribution scale (or *normalized*) to have a mean of 0 and standard deviation of 1, thereby following a normal distribution.

North American football

A synonym for American football.

odds

In betting and logistic regression, the number of times an event occurs in relation to the number of time the event does not occur. For example, if Kansas City has 4-to-1 odds of winning this week, we would expect them to win one game for every four games they lose under a similar situation. Odds can either emerge empirically through events occurring and models estimating the odds or through betting as odds emerge through the "wisdom" of the crowds.

odds-ratio

Odds in ratio format. For example, 3-to-2 odds can be written as 3:2 or 1.5 odds-ratios.

open source

Describes software where code must freely accessible (anybody can look at the code) and freely available (not cost money).

origination

The process by which an oddsmaker or a bettor sets the line.

outliers

Data points that are far away from another data point.

overfit

Describes a model for which too many parameters have been estimated compared to the amount of data, or a model that fits one situation too well and does not apply to other situations.

p-values

The probability of obtaining the observed test statistic, assuming the null hypothesis of no difference is true. These values are increasingly falling out of favor with professional statisticians because of their common misuse.

Pearson's correlation coefficient

A value from –1 to 1. A value of 1 means two groups are perfectly positively correlated, and as one increases, the other increases. A value of –1 means two groups are perfectly negatively correlated, and as one increases, the other decreases. A value of 0 means no correlation, and the values for one group do not have any relation to the values from another group.

pipe

To pass the outputs from one function directly to another function. See also *data pipeline*.

play-by-play (data)

The recorded results for each play of a football game. Often this data is "row poor," in that there are far more features (columns) than plays (rows).

principal component analysis (PCA)

A statistical tool for creating fewer independent features from a set of features.

probability

A number between 0 and 1 that describes the chance of an event occurring. Multiple definitions of probability exist, including the frequentist definition, which is the long-term average under similar conditions (such as flipping a coin), and Bayesian, which is the belief in an event occurring (such as betting markets).

probability distributions

A mathematical function that assigns a value (specifically, a probability) between 0 and 1 to an event occurring.

proposition (bets)

A type of bet focusing on a specific outcome occurring, such as who will score the first touchdown. Also called *prop* for short.

push

A game for which the outcome lands on the spread and the better is refunded their money.

Pythonistas

People who use the Python language.

quartile

A quarter of the data based on numerical ranking. By definition, data can be divided into four quartiles.

random-effect model

A model with coefficients that are assumed to come from a shared distribution.

regression

A type of statistical model that describes the relationship between one response variable (a simple linear regression) and one or more predictor variables (a multiple regression). Also a special type of linear model.

regression candidate

With regression, observations are expected to regress to the mean (average) value through time. For example, a player who has a good year this year would be reasonably expected to have a year closer to average next year, especially if the source of their good year is a relatively unstable, or *noisy*, statistic.

relative risk

A method for understanding the results from a Poisson regression, similar to the outputs from a logistic regression with odds-ratios.

residual

The difference between a model's predicted value for an observation and the actual value for an observation.

run yards over average (RYOE)

The number of running yards a player obtains compared to the value expected (or average) from a model given the play's situation.

sabermetrics

Quantitative analysis of baseball, named after the Society for American Baseball Research (SABR).

scatterplot

A type of plot that plots points on both axes.

scrape

To use computer programs to download data from websites (as in web scraping).

set the line

The process of the oddsmaker(s) creating the odds.

Simpson's paradox

A statistical phenomena whereby relationships between variables change based on different groupings using other variables.

slope

The change in a trend through time and often used to describe regression coefficients with continuous predictor variables.

spread (bet)

A betting market in American football that is the most popular and easy to understand. The spread is the point value meant to split outcomes in half over a large sample of games. This doesn't necessarily mean the sportsbook wants half of the bets on either side of the spread though.

stability

Within the context of this book, stability of an evaluation metric is the metric's ability to predict itself over a predetermined time frame. Also, see *stability analysis* and *sticky stats*.

stability analysis

The measurement of how well a metric or model output holds up through time. For example, with football, we would care about the stability of making predictions across seasons.

standard deviation

A measure of the spread, or dispersion, in a distribution.

standard error

A measure of the uncertainty around a distribution, given the uncertainty and sample size.

sticky stats

A term commonly used in fantasy football for numbers that are stable through time.

short pass

A pass typically less than 20 yards, although the actual threshold may vary (e.g., the first-down marker).

short-yardage back

A running back who tends to play when only a few (or "short") number of yards are required to obtain a first down or a touchdown.

supervised learning

A type of statistical and machine learning algorithm where people know the groups ahead of time and the algorithm may be trained on data.

three true outcomes

Baseball's first, and arguably most important, outcomes that can be modeled across area walks, strikeouts, and home runs. These outcomes also do not depend on the defense, other than rare exceptions.

total (bet)

A simple bet on whether the sum of the two teams' points goes over or under a specified amount.

Total (for a game) (bet)

The number of points expected by the betting market for a game.

unsupervised learning

A type of statistical and machine learning algorithm where people do not know the groups ahead of time.

useR

People who use the R language.

variable

Depending on context, two definitions are used in the book. First, observations can be variable. For example, pass yards might be highly variable for a quarterback, meaning the quarterback lacks consistency. Second, a model can be variable. For example, air yards might be a predictor variable for the response variable completion in a regression model.

vig

See *vigorish*.

vigorish

The house (casino, bookie, or other similar institution that takes bets) advantage that ensures the house almost always makes money over the long-term.

win probability (WP)

A model to predict the probability that a team wins the game at a given point during the game.

wins above replacement

A framework for estimating the number of wins a player is worth during the course of a season, set of seasons, or a career. First created in baseball.

yards per attempt (YPA)

Also known as *yards per passing attempt*, YPA is the average number of yards a

quarterback throws during a defined time period, such as game or season.

yards per carry (YPC)

The average number of yards a player runs the ball during a defined time period, such as game or season.

yards to go

The number of yards necessary to either obtain a first down or score during a play.

Index

floating-point numbers
    in Python, 266
    in R, 267
football analytics
    baseball analytics, compared, 3
    conferences for, 3
    history of, 1-3
    resources for information, 17, 259
    scouting, compared, 19-20
    tools for, 6-8
football markets, 138-141
for loops, 150-152
    in Python, 174-176
    in R, 179
forward pass, origin of, 4
4th Down Bot, 2
frequentist statistics, 242
full joins, 303-304
Fuller, Kendall, 52

G

Galton, Francis, 56
gambling (see sports betting)
gamma regression, 169
generalized additive mixed-effect models
    (GAMMs), 170
generalized additive models (GAMs), 170
generalized linear models (GLMs), 169-170
    building logistic regressions, 118-121
    completion percentage over expected,
        122-128
    purpose of, 117-118
    resources for information, 134, 171
geometric mean, 274
Gettleman, Dave, 5
ggplot2 tool, 31, 53
Git, 250-254
Git for Windows, 247
GitHub, 252
GitHub Copilot, 257
GitHub Desktop, 251
GitLab, 252
greater than (>) operator, 288
greater than or equal to (>=) operator, 288
Green Bay Packers
    averages in passing statistics, 272-275
    cheeseheads, 9
    drafting proficiency, 197, 200
    filtering data example, 291-296

home-field advantage example, xiv-xvi
grouping data
    passing statistics, 37-41
    rushing statistics, 69-72

H

Harris, Franco, 105
hashtag (#), 9
help files, 261
Henry, Derrick, 55, 71, 102, 105
Hermsmeyer, Josh, 5
hexbin plots, 120
The Hidden Game of Football (Carroll, Palmer,
    Thorn), 2
Hill, Taysom, 103, 125
histograms, 30-35
history of football analytics, 1-3
hold, 149
home-field advantage, xiv-xvi
Horowitz, Max, 2
house advantage, 5, 139

I

IDEs (integrated development environments),
    264-265
importing pandas package, 14
improving statistics presentations, 284
in operators, 289
independent variables (see predictor variables)
Indianapolis Colts, draft pick trade with New
    York Jets, 182, 192-194
individual player markets, 149-162
inner joins, 305
install.packages() function, 263
installing
    nflfastR package, 11
    packages, 11, 263-264
    Python, 261-262
    R, 261-262
    statsmodels package, 58
integers
    in Python, 265
    in R, 267
integrated development environments (IDEs),
    264-265
interactions, 95
interactives, 257
intercept, 80
intermediate objects versus piping objects, 35

interquartile range (IQR), 27, 275

J

Jackson, Bo, 4
Jackson, Lamar, 71, 103
Jacobs, Josh, 200
Jimmy Johnson chart, 183
Johnson, Calvin, 236
Johnson, Jimmy, 183
joins
    inner, 305
    left, 306
    outer (full), 303-304
    right, 305
Jones, Jerry, 107, 183
Jupyter Notebook, 257, 263

K

k-means clustering (see clustering)
kableExtra package, 99

L

Landry, Tom, 183
Lange, Gregg, 21
LaTeX, 257
left datasets, 303
left joins, 306
less than (<) operator, 288
less than or equal to (<=) operator, 288
line breaks, 29, 96, 296
linear regression, 117
    generalized linear models, 169-170
        building logistic regressions, 118-121
        completion percentage over expected,
            122-128
        purpose of, 117-118
    multiple
        assumption of linearity, 108-111
        defined, 80
        rushing yards over expected, 79-81,
            94-100
    simple
        defined, 56
        limitations of, 79
        rushing yards over expected, 56-58,
            64-69
        terminology, 56
link functions, 118

linting, 254
Linux, 251
lists
    in Python, 266
    in R, 268
loading
    nflfastR package, 11
    packages
        in Python, 22
        in R, 23
    play-by-play (pbp) data, 12, 22
local installation of Python/R, 261-262
log(0) function, 189
logic operators
    data wrangling, 287-291
    filtering data, 278-279
logical operators in R, 268
logistic regression, 118
    building, 118-121
    odds ratios, 132-134
logit, 118
lognormal regression, 169
long passes, 24
longitudinal data, 138
Lopez, Michael, 57, 183
ls command (bash), 248
Luck, Andrew, 182

M

machine learning, 246
Mack, Khalil, 200
macOS, plotting data, 31
Mahomes, Patrick, 41, 55, 131, 149-162, 194,
    273
markets in football betting, 138-141
Massey, Cade, 183
matplotlib package, 166
matrices, 268
McCoy, Mike, 183
mean, 272-275
median, 273
merging datasets, 302-308
metadata, 23, 178
Microsoft PowerShell, 247
MIT Sloan Sports Analytics Conference, 3
mode, 273
model matrix, 80
moneyline market, 139
Moss, Randy, 236

resources for information, 77, 112
simple linear
    defined, 56
    limitations of, 79
    rushing yards over expected, 56-58,
        64-69
    terminology, 56
regression candidates, 51
regression toward the mean, 159
relative paths, 249
relative risk, 169
reports, writing, 257
residuals
    assumption of linearity, 108-111
    in completion percentage over expected,
        131-132
    defined, 2
    multiple linear regression, 94
    simple linear regression, 67-68
resources for information
    football analytics, 17, 259
    generalized linear models, 134, 171
    NFL Scouting Combine, 203
    plotting data, 53
    probabilities, 171
    regression, 77, 112
    running back valuation, 77
    sports betting, 171
    statistics, 238, 285
    web scraping, 202
response variables, 56, 64
resumes (GitHub), 253
right datasets, 303
right joins, 305
right skew, 169
risk ratio, 169
Riske, Timo, 183, 200
rm command (bash), 249
Rodgers, Aaron, 43-45, 125
rolling average, 186
Roosevelt, Teddy, 4
running average, 186
running backs
    importance relative to other positions, 4-5,
        105-108
    resources for information, 77
    rushing statistics (see rushing yards over
        expected)
    short-yardage, 57

rushing yards over expected (RYOE)
    exploratory data analysis (EDA), 58-64,
        82-94
    grouping data, 69-72
    multiple linear regression, 79-81, 94-100
    simple linear regression, 56-58, 64-69
    stability analysis of, 73-76, 100-105
Ryan, Matt, 41, 52, 124, 128

S
S language, 69
sabermetrics, 3
Sanders, Barry, 4, 105
saving data files as outputs, 298
scaling, 221
scatterplots, 47-50
scikit-learn package, 69
scouting
    football analytics, compared, 19-20
    NFL Scouting Combine, 174, 203-204
scraping (see web scraping)
scripts, 262-263
seaborn package, 30, 53
sell-high candidates, 20
set the line, 137
Seth, Tej, 57
Shanahan, Kyle, 41
shaping (see data wrangling)
sharing data, 257
shell, 246
short passes, 24, 41-50
short-yardage backs, 57
shot quality, 2
significant digits, 273
Silver, Nate, 20
simple linear regression
    defined, 56
    limitations of, 79
    rushing yards over expected, 56-58, 64-69
    terminology, 56
Simpson's paradox, 86
single quotation marks ('), 11, 296
skill positions, 233
slope, 80
smart quotes, 263
Smith, Alex, 52
Smith, Emmitt, 4, 107
smoothing noise, 186

W

Walker, Herschel, 4, 183
Walsh, Bill, 1
Watson, Deshaun, 124, 128
web scraping
    NFL Scouting Combine data, 205-217
    with Python, 174-179
    with R, 179-182
    resources for information, 202
    terms and conditions, 175
West Coast offense, 1
white space, 29, 296
Wickham, Hadley, 270, 287
wide receivers, 236
win probability (WP), 2

window, 186
Windows Subsystem for Linux (WSL), 247
wins above replacement (WAR), 2
Winston, Jameis, 41, 128
wisdom of crowds, 137
wrangling data (see data wrangling)
writing reports, 257

Y

yards to go, 57
Yurko, Ron, 2

Z

Zsh shell, 247

About the Authors

Eric A. Eager is the vice president and partner at SumerSports, a football analytics startup founded by billionaire hedge fund manager Paul Tudor Jones and his son, Jack Jones. Eric currently hosts the *SumerSports Show* with former Falcons General Manager and two-time NFL Executive of the Year Thomas Dimitroff. Prior to joining Sumer, he founded the industry-leading analytics group at Pro Football Focus (PFF), which is owned by former Bengals wide receiver and current Sunday Night Football color commentator, Cris Collinsworth.

During his career, Eric has built tools used by all 32 NFL teams, over 100 college football teams, and various media entities. His simulation model has been used by NBC's Steve Kornacki during his "Road to the Playoffs" segment on NBC since 2020; he was also a part of the Fox NFL Game of the Week broadcast with Joe Buck and Troy Aikman, before they left for ESPN in 2022. His former podcast, the *PFF Forecast*, was the most popular football analytics podcast in the world when he left the show.

Eric studied applied mathematics and mathematical biology at the University of Nebraska–Lincoln, where he wrote his PhD thesis on how stochasticity and nonlinear processes affect population dynamics. For the first six years of his career, he was a professor at the University of Wisconsin La Crosse, where he published more than 25 peer-reviewed research papers on the interface of mathematics, biology, and the scholarship of teaching and learning. Six of those papers were published with Richard Erickson, his coauthor for this book. After leaving academia for football full-time in 2018, he's maintained a connection to teaching by building "Linear Algebra for Data Science in R" for DataCamp in 2018, as well as teaching Wharton's Moneyball Academy course to high school students each summer since 2020.

Eric maintains a strong interest in mentorship, as he wants up-and-coming football analysts to have a more straightforward path to a career than he had. He enjoys reading, writing, biking, rowing, and watching the WNBA with his family. He lives in Atlanta, Georgia, with his wife, Stephanie, and daughters Madeline and Chloe.

Richard A. Erickson helps people use mathematics and statistics to understand our world as well as make decisions with this data. He is a lifelong Green Bay Packers fan, and, like thousands of other Cheeseheads, a team owner. He has taught over 32,000 students statistics through graduate-level courses, workshops, and his Data-Camp courses "Generalized Linear Models in R" and "Hierarchical and Mixed Effects Models in R." He also uses Python on a regular basis to model scientific problems.

Richard received his PhD in environmental toxicology with an applied math minor from Texas Tech where he wrote his dissertation on modeling population-level effects of pesticides. He has modeled and analyzed diverse datasets including topics such as soil productivity for the US Department of Agriculture, impacts of climate change on disease dynamics, and improving rural healthcare. Richard currently works as a research scientist and has over 80 peer-reviewed publications. Besides teaching Eric about R and Python, Richard also taught Eric to like cheese curds.

Richard lives in La Crosse, Wisconsin, with his daughter, Margo, and Bernese Mountain Dog, Sadie. When not cheering for his Packers, he likes silent sports, notably biking, hiking, cross-country skiing, sea kayaking, and scuba diving.

Colophon

The animal on the cover of *Football Analytics with Python & R* is a white-crested laughingthrush. Its scientific name, *Garrulax leucolophus*, comes from the Latin term *garrire*, which means "to chatter," and from the Greek words *leukós* (white) and *lophos* (crest), in reference to their white hoods and raised crests.

As their scientific name suggests, white-crested laughingthrushes are very vocal birds that live in the forest regions throughout the Himalayas and Southeast Asia. They are highly social by nature and can travel in flocks of up to forty birds. Their leader initiates short (but loud) cackling calls, which are often answered with a chorus; while their calls can be jarring at first, they're often followed by quieter, more pleasant chatter. They favor dense thickets and shrubs, where they can hide from predators and seek shelter from the sun. These birds eat a variety of insects, fruits, seeds, and even small reptiles.

The cover illustration is by Karen Montgomery, based on an antique line engraving from *Shaw's Zoology*. The cover fonts are Gilroy Semibold and Guardian Sans. The text font is Adobe Minion Pro; the heading font is Adobe Myriad Condensed; and the code font is Dalton Maag's Ubuntu Mono.

O'REILLY®

Learn from experts.
Become one yourself.

Books | Live online courses
Instant answers | Virtual events
Videos | Interactive learning

Get started at oreilly.com.

Printed in the USA
CPSIA information can be obtained
at www.ICGtesting.com
JSHW060038180624
64967JS00014B/340

9 781492 099628